13

CRM SERIES

Centro
di Ricerca
Matematica
Ennio De Giorgi

Bruno De Maria
Dipartimento di Matematica e Applicazioni
Complesso Universitario di Monte Sant'Angelo
Via Cintia
80126 Napoli, Italia
b.demaria@unina.it

Nicola Fusco
Dipartimento di Matematica e Applicazioni
Complesso Universitario di Monte Sant'Angelo
Via Cintia
80126 Napoli, Italia
n.fusco@unina.it

Ernst Kuwert
Mathematisches Intitut
der Albert-Ludwigs-Universität Freiburg
Eckerstraße 1
D-79104 Freiburg, Germany
kuwert@mathematik.uni-freiburg.de

Tristan Rivière
Department of Mathematics
ETH Zentrum
CH-8093 Zürich, Switzerland
tristan.riviere@firm.math.ethz.ch

Reiner Schätzle
Mathematisches Institut
der Eberhard-Karls-Universität Tübingen
Auf der Morgenstelle 10
D-72076 Tübingen, Germany
schaetz@everest.mathematik.uni-tuebingen.de

Topics in Modern Regularity Theory

edited by
Giuseppe Mingione

EDIZIONI
DELLA
NORMALE

ISBN: 978-88-7642-426-7
e-ISBN 978-88-7642-427-4

Contents

Introduction

This volume collects the contributions of some of the leading experts in PDE who gave courses in the two intensive research periods I organized at the Centro De Giorgi of Scuola Normale Superiore at Pisa and at the University of Parma, in September 2009 and in the Spring of 2010, respectively. The speakers kindly agreed to give courses whose aims were both to review the established theory and to present the latest research developments; the notes included here summarize and extend the content of the lectures given in some of the courses offered by the schools. Specifically, the book contains three different contributions: the first one, in order of presentation, is by Ernst Kuwert & Reiner Schätzle, the second by Tristan Rivière, the third and final one by Bruno De Maria & Nicola Fusco. The first two parts are of expository character, and summarize some recent results obtained by the authors, after giving a rather general and comprehensive introduction to the subject. The third one contains some new results together with an up-to-dated presentation of the setting of problems dealt with.

I hereby take the opportunity to acknowledge the support of the European Research Council via the ERC Grant 207573 "Vectorial problems" and to thank the colleagues who were also responsible of the organization of the intensive periods, and, amongst them, especially Frank Duzaar and Juha Kinnunen.

The volume starts with the beautiful lecture notes, simply titled "The Willmore functional", by Ernst Kuwert & Reiner Schätzle. They give a very comprehensive introduction to the basic analytic aspects of the analysis of Willmore surfaces, *i.e.* the critical points of the Willmore functional, smoothly taking the reader from the basic facts to some of the most updated current research issues. After recalling the starting definitions and introducing a number of related tools, such as for instance monotonicity formulas, the authors present a careful analysis of basic aspects of the Willmore flow (the gradient flow associated to the Willmore functional) such as estimates on the maximal existence time interval and then the blow-up analysis of singularities at the time of their formation; asymptotic convergence properties are considered and studied as well. The authors then proceed in the analysis of the conformal parametrization properties of surfaces; it is here remarkable to note how Kuwert & Schätzle succeed in giving a smooth, clear and self-contained of some certainly not easy pieces of work, as for instance the study of asymptotic properties of the classical conformal parametrization of Huber made by Müller & Šverák a few years ago. Further topics treated in the notes are concerned with the removability of point singularities and applications to global existence of the Willmore flow of embedded surfaces, that the authors present in connection with the results of Robert Bryant; further connections emerge here with recent work of Rivière, partially related to the content of the subsequent chapter of this book. Finally, the authors give proofs of basic theorems in the variational analysis of Willmore functional such as those concerning compactness via the Moebius group quotients, and minimization asymptotic problems in classes of surfaces with prescribed genus.

In his "The role of conservation laws in the analysis of conformally invariant problems", Tristan Rivière gives a very comprehensive and updated presentation of regularity techniques aimed at treating conformally invariant variational problems. This is a longstanding and traditional topic in the modern Calculus of Variations. Rivière reviews a number of basic relevant techniques – as for instance compensated compactness and integrability by compensation – and then proceeds to explain the use of conservation laws in the regularity analysis of certain systems with critical growth. Finally, he explains his proof of the famous Hildebrandt's conjecture. This states the Hölder continuity of energy critical points of conformally invariant quadratic growth functionals. Rivière's proof is based on the new observation that the regularity of this problem can be treated by analyzing certain systems with antisymmetric potential; in turn Rivière's analysis identifies the central role of antisymmetry of potentials in allowing for deriving conservation laws when treating the regularity of systems with critical growth right hand side. This new and groundbreaking approach is robust enough to allow for many other applications to conformally invariant problems and in Geometric Analysis. In the last part of his notes, yet a new and surprising approach is described: the discovery of the robustness of the traditional ODE method of the variation of constants in the setting of Schrödinger systems with anti-symmetric potentials. Indeed, a new formulation of this approach is proposed and shown to be an effective tool when dealing with the regularity of more general critical growth right hand side systems.

De Maria & Fusco present a contribution titled "Regularity properties of equilibrium configurations of epitaxially strained elastic films" which is devoted to the mathematical study of the morphological instabilities of interfaces generated by the competition between elastic energy and surface tension, the so-called stress driven rearrangement instabilities (SDRI) of surfaces and interfaces in solids. Such a topic has been the focus of a rapidly growing interest of the applied and computational communities, also in view of its several important technological applications. Besides this, their paper can be also seen as a fine contribution to the regularity theory of the so-called free-discontinuity problems. Morphological instabilities occur, for instance, in the hetero-epitaxial growth of thin films for systems with a lattice mismatch between film and substrate. When the film is grown on a flat substrate, its profile remains flat until a critical value of the thickness is reached, after which the free surface develops corrugations, material clusters, and, possibly, cusp singularities. This is commonly referred to as the Asaro-Grinfeld-Tiller (AGT) instability, after the name of the scientists who started such theoretical investigations. Several numerical and theoretical studies have been carried out to study quantitative and qualitative properties of equilibrium configurations of strained epitaxial films. Although very insightful, most of these works lack rigorous mathematical content. Eventually, the foundations of a rigorous mathematical treatment have been given in works by Grinfeld (Soviet Physics Doklady, 1986), Bonnetier & Chambolle (SIAM J. Appl. Math., 2002), and more recently in work by Fonseca, Fusco, Leoni and Morini (ARMA 2007), who developed a complete regularity theory for a variant of the Bonnetier-Chambolle functional, modeling the case of an infinitely thick elastic substrate. Following the path set by Fonseca, Fusco, Leoni and Morini, De Maria & Fusco extend all these results to the functional originally considered by Bonnetier & Chambolle, which deals with the case of a rigid substrate. The main technical achievement is the rigorous validation of zero contact angle condition. The proof of this fact turns out to be considerably more difficult than in the case considered by Fonseca, Fusco, Leoni and Morini, since the presence of a Dirichlet condition at the interface between film and substrate poses non-trivial additional difficulties. The regularity results established by De Maria & Fusco presented here have been in fact used in several subsequent papers.

Parma, October 2011

Giuseppe Mingione

The Willmore functional

Ernst Kuwert and Reiner Schätzle

Contents **1**

E. Kuwert and R. Schätzle were supported by the DFG Sonderforschungsbereich TR 71 Freiburg - Tübingen.

1 Introduction to Geometry

1.1 Introduction

For an immersed closed surface $f : \Sigma \to \mathbb{R}^n$ the Willmore functional is defined by

$$\mathcal{W}(f) = \frac{1}{4} \int_\Sigma |\vec{\mathbf{H}}|^2 \, d\mu_g.$$

Here $g = f^* g_{\mathrm{euc}}$ denotes the pull-back metric of the Euclidean metric under f, that is in local coordinates

$$g_{ij} := \langle \partial_i f, \partial_j f \rangle.$$

Moreover, $g = \det(g_{ij})$, $(g^{ij}) = (g_{ij})^{-1}$ and for the induced area measure

$$\mu_f = \mu_g = \sqrt{g}\, \mathcal{L}^2.$$

The second fundamental form of f is the normal projection of the second derivatives of f

$$A_{ij} := (\partial_{ij} f)^\perp.$$

We define the mean curvature vector and the tracefree second fundamental form by

$$\vec{\mathbf{H}} = g^{ij} A_{ij} \quad \text{and} \quad A_{ij}^0 = A_{ij} - \frac{1}{2} \vec{\mathbf{H}} g_{ij}.$$

The Gauß curvature can be written by the Gauß equations, see [dC, Section 6, Proposition 3.1], as

$$K = \langle A(e_1, e_1), A(e_2, e_2)\rangle - \langle A(e_1, e_2), A(e_1, e_2)\rangle, \tag{1.1.1}$$

and combining with the inequality of geometric and arithmetic mean

$$|K| \leq |A|^2/2. \tag{1.1.2}$$

In any orthonormal basis e_1, e_2 of the tangent space, we see

$$\begin{aligned}
\vec{\mathbf{H}} &= A(e_1, e_1) + A(e_2, e_2),\\
|A|^2 &= |A(e_1, e_1)|^2 + |A(e_1, e_2)|^2 + |A(e_2, e_1)|^2 + |A(e_2, e_2)|^2.
\end{aligned}$$

We calculate

$$\begin{aligned}
|\vec{\mathbf{H}}|^2 &= |A(e_1, e_1)|^2 + 2\langle A(e_1, e_1), A(e_2, e_2)\rangle + |A(e_2, e_2)|^2\\
&= |A|^2 + 2\big(\langle A(e_1, e_1), A(e_2, e_2)\rangle - \langle A(e_1, e_2), A(e_1, e_2)\rangle\big),
\end{aligned} \tag{1.1.3}$$

hence by (1.1.1)

$$|\vec{\mathbf{H}}|^2 = |A|^2 + 2K. \tag{1.1.4}$$

Likewise

$$\begin{aligned}
\frac{1}{2}|A^0|^2 &= \left|\frac{A(e_1, e_1) - A(e_2, e_2)}{2}\right|^2 + |A(e_1, e_2)|^2\\
&= \frac{1}{4}|A(e_1, e_1)|^2 + \frac{1}{4}|A(e_2, e_2)|^2 + \frac{1}{2}|A(e_1, e_2)|^2\\
&\quad - \frac{1}{2}\Big(\langle A(e_1, e_1), A(e_2, e_2)\rangle - \langle A(e_1, e_2), A(e_1, e_2)\rangle\Big)\\
&= \frac{1}{4}|A|^2 - \frac{1}{2}\Big(\langle A(e_1, e_1), A(e_2, e_2)\rangle - \langle A(e_1, e_2), A(e_1, e_2)\rangle\Big)
\end{aligned}$$

and using (1.1.3)

$$\frac{1}{2}|A^0|^2 = \frac{1}{4}|\vec{\mathbf{H}}|^2 - \langle A(e_1, e_1), A(e_2, e_2)\rangle + \langle A(e_1, e_2), A(e_1, e_2)\rangle. \tag{1.1.5}$$

Again by (1.1.1), we obtain

$$\frac{1}{4}|\vec{\mathbf{H}}|^2 - K = \frac{1}{2}|A^0|^2 \tag{1.1.6}$$

and combining with (1.1.4)

$$|A|^2 = 2|A^0|^2 + 2K. \tag{1.1.7}$$

For a closed surface Σ the integral over the Gauß curvature is given by the Gauß-Bonnet theorem as

$$\int_\Sigma K \, d\mu_g = 2\pi(\Sigma), \tag{1.1.8}$$

where $\chi(\Sigma)$ denotes the Euler characteristic of Σ. This yields with (1.1.4) and (1.1.6)

$$\mathcal{W}(f) = \frac{1}{4}\int_\Sigma |A|^2 \, d\mu_g + \pi\chi(\Sigma) = \frac{1}{2}\int_\Sigma |A^\circ|^2 \, d\mu_g + 2\pi\chi(\Sigma). \tag{1.1.9}$$

We finish this introduction establishing a lower bound in codimension one.

Proposition 1.1.1 ([Wil65]). *For any embedding $f : \Sigma \to \mathbb{R}^3$ of a closed surface Σ, we have*

$$\mathcal{W}(f) \geq 4\pi$$

and equality implies that f parametrises a round sphere.

Proof. We consider $\Sigma \subseteq \mathbb{R}^3$. Let $\nu : \Sigma \to S^2$ be the unique smooth outer normal at Σ. For any unit vector ν_0, we choose $x_0 \in \Sigma$ with

$$\langle x_0, \nu_0 \rangle := \max_{x \in \Sigma} \langle x, \nu_0 \rangle$$

and see

$$\Sigma \subseteq \{ y \in \mathbb{R}^3 \mid \langle y - x_0, \nu_0 \rangle \leq 0 \}.$$

Therefore $\{\nu_0\}^\perp$ is a supporting hyperplane of Σ at x_0 and

$$\nu(x_0) = \nu_0 \quad \text{and} \quad K(x_0) \geq 0.$$

As $\nu_0 \in S^2$ was arbitrary, we get

$$\nu(K \geq 0) = S^2. \tag{1.1.10}$$

We define the scalar second fundamental form

$$h_{ij} = \langle A_{ij}, \nu \rangle = \langle \partial_{ij} f, \nu \rangle.$$

Clearly, the Gauß curvature is the determinant

$$K = \det{}_g(h_{ij}).$$

As $\partial f \perp \nu$, we get

$$h_{ij} = -\langle \partial_i \nu, \partial_j f \rangle.$$

Considering $g_{ij} = \delta_{ij}$ at some point, we get that $\partial_1 f, \partial_2 f, \nu$ is an orthonormal basis of $\mathbb{R}^3$ and observing $\partial \nu \perp \nu$, as $|\nu| = 1$,

$$\partial_i \nu = -h_{i1} \partial_1 f - h_{i2} \partial_2 f.$$

We calculate the Jacobian

$$(J_g \nu)^2 = \det(\langle \partial_i \nu, \partial_j \nu \rangle) = \det(h_{ij})^2 = K^2,$$

hence

$$J_g \nu = |K|.$$

By (1.1.6), (1.1.10) and the area formula, we get

$$\mathcal{W}(\Sigma) \geq \int_{[K>0]} K \, \mathrm{d}\mu_g = \int_{[K \geq 0]} J_g \nu \, \mathrm{d}\mu_g \geq \mathcal{H}^2(\nu(K \geq 0)) = \mathcal{H}^2(S^2) = 4\pi,$$

in particular $[K > 0] \neq \emptyset$. In case of equality, we see using (1.1.10)

$$0 = \int_{[K>0]} \left(\frac{1}{4} |\vec{\mathbf{H}}|^2 - K \right) \mathrm{d}\mu_g = \int_{[K>0]} \frac{1}{2} |A^0|^2 \, \mathrm{d}\mu_g,$$

hence $A^0 \equiv 0$ in $[K > 0]$. By a theorem of Codazzi, f parametrises in any connected component Ω of $[K > 0]$ a piece of a round sphere $\partial B_R(a)$, in particular $K \equiv 1/R^2$ is constant in Ω. Therefore Ω is closed, hence $\Omega = [K > 0] = \Sigma$ and $A^0 \equiv 0$ on Σ, as Σ is connected and $[K > 0] \neq \emptyset$. Then $f : \Sigma \to \partial B_R(a)$ is a local diffeomorphism, hence a covering map, and, as $\partial B_R(a)$ is simply connected, $f : \Sigma \xrightarrow{\approx} \partial B_R(a)$ is a diffeomorphism. □

1.2 Conformal invariance

Clearly, the Willmore functional is invariant under isometries. Oberserving for $\lambda > 0$

$$\vec{\mathbf{H}}_{\lambda f} = \lambda^{-1} \vec{\mathbf{H}} \quad \text{and} \quad \mu_{\lambda f} = \lambda^2 \mu_f,$$

we see

$$\mathcal{W}(\lambda f) = \mathcal{W}(f),$$

and the Willmore functional is also invaraint under homotheties.

More general the Willmore functional is invariant under conformal transformations. We start with the following pointwise invariance.

Proposition 1.2.1 ([Ch74]). *Let* M *be a* n*-dimensional manifold with two conformal metrics* $\bar{g} = e^{2u} g$ *and* $\Sigma \subseteq M$ *be a* m*-dimensional submanifold. Then the second fundamental forms* A *and* $\bar{A}$ *of* Σ *with respect to the ambient metrics* g *and* $\bar{g}$ *satisfy*

$$\bar{A}_{ij} - A_{ij} = -g_{ij} \operatorname{grad}_g^{\perp} u \tag{1.2.1}$$

in local charts on Σ*, where* $\operatorname{grad}_g u^k = g^{kl} \partial_l u$ *denotes the gradient of* u *with respect to* g *and* $.^{\perp}$ *denotes the component normal to* Σ *with respect to either* g *or* $\bar{g}$*. In particular*

$$\bar{A}^0_{ij} = A^0_{ij} \tag{1.2.2}$$

for the tracefree second fundamental forms and for a surface Σ *that is* $m = 2$

$$|\bar{A}^0|^2_{\bar{g}}\, \mu_{\bar{g}} = |A^0|^2_g\, \mu_g. \tag{1.2.3}$$

First we compute the difference of Christoffel symbols for conformal metrics.

Proposition 1.2.2. *Let* M *be a* n*-dimensional manifold with two conformal metrics* $\bar{g} = e^{2u} g$*. The difference of the Christoffel symbols in local coordinates is a tensor and is given by*

$$T^k_{ij} := \bar{\Gamma}^k_{ij} - \Gamma^k_{ij} = \delta^k_i \partial_j u + \delta^k_j \partial_i u - g_{ij} g^{kl} \partial_l u \tag{1.2.4}$$

and in conformal coordinate $g_{ij} = e^{2v} \delta_{ij}$

$$\begin{aligned} T^1_{11} &= \partial_1 u, & T^1_{12} &= T^1_{21} = \partial_2 u, & T^1_{22} &= -\partial_1 u, \\ T^2_{11} &= -\partial_2 u, & T^2_{12} &= T^2_{21} = \partial_1 u, & T^2_{22} &= \partial_2 u. \end{aligned} \tag{1.2.5}$$

Proof. We calculate

$$\begin{aligned} 2\bar{\Gamma}^k_{ij} &= \bar{g}^{kl}(\partial_i \bar{g}_{jl} + \partial_j \bar{g}_{li} - \partial_l \bar{g}_{ij}) \\ &= g^{kl}(\partial_i g_{jl} + \partial_j g_{li} - \partial_l g_{ij}) \\ &\quad + e^{-2u} g^{kl}(g_{jl}\partial_i e^{2u} + g_{li}\partial_j e^{2u} - g_{ij}\partial_l e^{2u}) \\ &= 2\Gamma^k_{ij} + 2g^{kl}(g_{jl}\partial_i u + g_{li}\partial_j u - g_{ij}\partial_l u) \\ &= 2\Gamma^k_{ij} + 2(\delta^k_j \partial_i u + \delta^k_i \partial_j u - g^{kl} g_{ij} \partial_l u), \end{aligned}$$

which is (1.2.4), and (1.2.5) follows easily by direct evaluation. □

Proof of Proposition 1.2.1. Let Γ_{ij}^k and $\bar{\Gamma}_{ij}^k$ denote the Christoffel symbols of g and $\bar{g}$ and put $T_{ij}^k := \bar{\Gamma}_{ij}^k - \Gamma_{ij}^k$ which is a tensor. We calculate the covariant derivatives of a vectorfield $X = X^k \partial_k$ on M as

$$\nabla_i^{\bar{g}} X^k = \nabla_i^g X^k + T_{il}^k X^l.$$

We may consider $\Sigma = B_1^m(0) \times \{0\} \subseteq B_1^n(0) = M$ and $g_{ij}(0) = \delta_{ij}$. Then $\partial_1, \ldots, \partial_m$ are tangential to Σ and $\partial_{m+1}, \ldots, \partial_n$ form a basis of $T_0^\perp \Sigma$. Choosing $X = \partial_j, i, j = 1, \ldots, m$, we get for the normal projections of the covariant derivatives which are the second fundamental form

$$\bar{A}_{ij} - A_{ij} = \nabla_i^{\bar{g},\perp} \partial_j - \nabla_i^{g,\perp} \partial_j = T_{ij}^k \partial_k^\perp = \sum_{k=m+1}^n T_{ij}^k \partial_k. \qquad (1.2.6)$$

Using (1.2.4) and observing that $i, j = 1, \ldots, m, k = m+1, \ldots, n$ yields $g_{ik}(0) = g_{jk}(0) = 0$, as $g_{rs}(0) = \delta_{rs}$, hence

$$T_{ij}^k = -g_{ij} g^{kl} \partial_l u \quad \text{in } 0.$$

Plugging into (1.2.6), we obtain

$$\bar{A}_{ij} - A_{ij} = -\sum_{k=m+1}^n g_{ij} g^{kl} \partial_l u \partial_k = -g_{ij} g^{kl} \partial_l u \partial_k^\perp = -g_{ij} \operatorname{grad}_g^\perp u \quad \text{in } 0.$$

This equation is tensorial, and we obtain (1.2.1) on Σ. As the difference is a multiple of either metric g or $\bar{g}$, the tracefree parts coincide which is (1.2.2). Finally

$$\begin{aligned} |\bar{A}^0|_{\bar{g}}^2 \sqrt{\bar{g}} &= \bar{g}^{ik} \bar{g}^{jl} \bar{g}(\bar{A}_{ij}^0, \bar{A}_{kl}^0) \sqrt{\bar{g}} \\ &= e^{-4u} g^{ik} g^{jl} e^{2u} g(A_{ij}^0, A_{kl}^0) \sqrt{\det(e^{2u} g_{ij})} = |A^0|_g^2 \sqrt{g} \end{aligned}$$

which yields (1.2.3). □

We consider an immersion $f : \Sigma \to \Omega$ of a closed surface into an open set $\Omega \subseteq \mathbb{R}^n$. Let $\Phi : \Omega \xrightarrow{\approx} \Omega' \subseteq \mathbb{R}^n$ be a conformal diffeomorphism with pull-back metric $\bar{g} = \Phi^* g_{\text{euc}} = e^{2u} g_{\text{euc}}$. We calculate with (1.1.9) and (1.2.3)

$$\begin{aligned} \mathcal{W}(\Phi \circ f) &= \frac{1}{2} \int_\Sigma |A_{\Phi \circ f}^0|^2 \, d\mu_{\Phi \circ f} + 2\pi \chi(\Sigma) \\ &= \frac{1}{2} \int_\Sigma |A_f^0|_{\bar{g}}^2 \, d\mu_{\bar{g}} + 2\pi \chi(\Sigma) \\ &= \frac{1}{2} \int_\Sigma |A_f^0|^2 \, d\mu_f + 2\pi \chi(\Sigma) = \mathcal{W}(f), \end{aligned}$$

and the Willmore functional is invariant under conformal diffeomorphisms.

Now we want to extend the definition of the Willmore functional from Euclidean target to general n-dimensional manifold M with metric g, see [Wei78]. We consider an immersion $f : \Sigma \to M$ of a closed surface. The Gauß equations (1.1.1) extend in M, see [dC, Section 6, Proposition 3.1], using the Riemann curvature tensors R_Σ and R_M of Σ and M to

$$\begin{aligned}&\langle A(e_1, e_1), A(e_2, e_2)\rangle - \langle A(e_1, e_2), A(e_1, e_2)\rangle \\ &\quad = R_\Sigma(e_1, e_2, e_1, e_2) - R_M(e_1, e_2, e_1, e_2) = K_\Sigma - K_M^\Sigma,\end{aligned}$$

where K_Σ is Gauß curvature of Σ and K_M^Σ is the sectional curvature of M with respect to the tangent space of Σ. Recalling (1.1.5), which holds true in general M, we obtain

$$\frac{1}{4}|\vec{\mathbf{H}}|^2 + K_M^\Sigma = \frac{1}{2}|A^0|^2 + K_\Sigma.$$

The integral over Σ with respect to the area measure μ_g of the first term on the right hand side is a conformal invariant by Proposition 1.2.1, and the integral over the second term is a topological invariant. This yields the following definition and proposition.

Proposition 1.2.3. *For a immersion $f : \Sigma \to M$ of a closed surface Σ into a n-dimensional manifolds M with metric g, we define the Willmore functional*

$$\mathcal{W}(f) = \mathcal{W}(f, g) := \int_\Sigma \left(\frac{1}{4}|\vec{\mathbf{H}}|^2 + K_M^\Sigma\right) \mathrm{d}\mu_g. \tag{1.2.7}$$

The Willmore functional is invariant under conformal changes of the metric, that is

$$\mathcal{W}(f, \bar{g}) = \mathcal{W}(f)$$

for any conformal metric $\bar{g} = e^{2u} g$.

For the special case of a sphere $M = S^n$ with canonical metric, we have $K_{S^n}^\Sigma \equiv 1$ and get for $f : \Sigma \to S^n$ that

$$\mathcal{W}(f) = \frac{1}{4}\int_\Sigma |\vec{\mathbf{H}}|^2 \,\mathrm{d}\mu_g + \mathrm{Area}(f). \tag{1.2.8}$$

1.3 The Euler Lagrange equation

Critical points of the Willmore functional are called Willmore immersions or Willmore surfaces. Here we derive the Euler Lagrange equation for the Willmore functional.

We consider a smooth one-parameter family of immersions $f_t : \Sigma \to \mathbb{R}^n$ with $\partial_t f_{t,|t=0} = V$ normal along f. We get in local coordinates

$$g_{t,ij} = \langle \partial_i f_t, \partial_j f_t \rangle$$

and

$$\partial_t g_{ij} = \langle \partial_i f, \partial_j V \rangle + \langle \partial_j f, \partial_i V \rangle = -2\langle A_{ij}, V \rangle, \tag{1.3.1}$$

as $\partial f \perp V$ and $A = \partial^2 f^\perp$. Then as $g^{ij} g_{jk} = \delta^i_k$

$$\partial_t g^{ij} = -g^{ik} \partial_t g_{kl} g^{lj} = 2 g^{ik} g^{jl} \langle A_{kl}, V \rangle \tag{1.3.2}$$

and

$$\partial_t g = \partial_t \det(g^{ij}_{t=0} g_{jk}) g = \mathrm{tr}(g^{ij} \partial_t g_{jk}) g = -2 g^{ij} \langle A_{ij}, V \rangle g = -2 \langle \vec{\mathbf{H}}, V \rangle g,$$

hence

$$\partial_t \mu_g = \partial_t \sqrt{g} \mathcal{L}^2 = -\langle \vec{\mathbf{H}}, V \rangle \sqrt{g} \mathcal{L}^2 = -\langle \vec{\mathbf{H}}, V \rangle \mu_g. \tag{1.3.3}$$

Next we write ∇_i for the covariant derivative, $\nabla_i^\perp$ for its normal projection and $\partial_t^\perp$ for the normal projection of the time derivative. We recall the Weingarten equations

$$\begin{aligned} A_{ij} &= (\partial_{ij} f)^\perp = \partial_{ij} f - \langle \partial_{ij} f, \partial_k f \rangle g^{kl} \partial_l f \\ &= \partial_{ij} f - \Gamma^k_{ij} \partial_k f = \nabla_i \nabla_j f. \end{aligned}$$

We calculate

$$\begin{aligned} \partial_t^\perp A_{ij} &= (\partial_{ij} V)^\perp - \langle \partial_{ij} f, \partial_k f \rangle g^{kl} \partial_t^\perp \partial_l f \\ &= (\partial_{ij} V - \Gamma^k_{ij} \partial_k V)^\perp = (\nabla_i \nabla_j V)^\perp \\ &= \nabla_i^\perp \nabla_j^\perp V + \nabla_i^\perp \Big(\langle \partial_j V, \partial_k f \rangle g^{kl} \partial_l f \Big) \\ &= \nabla_i^\perp \nabla_j^\perp V - \langle A_{jk}, V \rangle g^{kl} \nabla_i^\perp \partial_l f \\ &= \nabla_i^\perp \nabla_j^\perp V - \langle A_{jk}, V \rangle g^{kl} A_{il} \end{aligned} \tag{1.3.4}$$

and

$$\begin{aligned}
\partial_t^\perp \vec{\mathbf{H}} = \partial_t^\perp (g^{ij} A_{ij}) &= g^{ij} \Big(\nabla_i^\perp \nabla_j^\perp V - \langle A_{jk}, V \rangle g^{kl} A_{il} \Big) \\
&\quad + 2 g^{ik} g^{jl} \langle A_{kl}, V \rangle A_{ij} \\
&= \Delta^\perp V + g^{ik} g^{jl} \langle A_{ij}, V \rangle A_{kl} \\
&= \Delta^\perp V + g^{ik} g^{jl} \Big\langle A_{ij}^0 + \frac{1}{2} \vec{\mathbf{H}} g_{ij}, V \Big\rangle \Big(A_{kl}^0 + \frac{1}{2} \vec{\mathbf{H}} g_{kl} \Big) \\
&= \Delta^\perp V + g^{ik} g^{jl} \langle A_{ij}^0, V \rangle A_{kl}^0 \\
&\quad + \frac{1}{2} g^{ik} g^{jl} g_{ij} A_{kl}^0 \langle \vec{\mathbf{H}}, V \rangle + \frac{1}{2} g^{ik} g^{jl} g_{kl} \langle A_{ij}^0, V \rangle \vec{\mathbf{H}} \\
&\quad + \frac{1}{4} g^{ik} g^{jl} g_{ij} g_{kl} \langle \vec{\mathbf{H}}, V \rangle \vec{\mathbf{H}} \\
&= \Delta^\perp V + g^{ik} g^{jl} \langle A_{ij}^0, V \rangle A_{kl}^0 + \frac{1}{2} \langle \vec{\mathbf{H}}, V \rangle \vec{\mathbf{H}},
\end{aligned}$$

where $\Delta^\perp$ denotes the Laplacian in the normal bundle. Combining we get

$$\begin{aligned}
\partial_t \Big(|\vec{\mathbf{H}}|^2 \mu_g \Big) &= 2 \langle \partial_t^\perp \vec{\mathbf{H}}, \vec{\mathbf{H}} \rangle \mu_g + |\vec{\mathbf{H}}|^2 \partial_t \mu_g \\
&= \Big(2 \langle \Delta^\perp V, \vec{\mathbf{H}} \rangle + 2 g^{ik} g^{jl} \langle A_{ij}^0, \vec{\mathbf{H}} \rangle \langle A_{kl}^0, V \rangle \\
&\qquad + \langle \vec{\mathbf{H}}, V \rangle |\vec{\mathbf{H}}|^2 - |\vec{\mathbf{H}}|^2 \langle \vec{\mathbf{H}}, V \rangle \Big) \mu_g \\
&= 2 \Big(\langle \Delta^\perp V, \vec{\mathbf{H}} \rangle + g^{ik} g^{jl} \langle A_{ij}^0, \vec{\mathbf{H}} \rangle \langle A_{kl}^0, V \rangle \Big) \mu_g .
\end{aligned} \tag{1.3.5}$$

Integrating yields

$$\begin{aligned}
\frac{d}{dt} \mathcal{W}(f_t) &= \frac{1}{2} \int_\Sigma \Big(\langle \Delta^\perp V, \vec{\mathbf{H}} \rangle + g^{ik} g^{jl} \langle A_{ij}^0, \vec{\mathbf{H}} \rangle \langle A_{kl}^0, V \rangle \Big) \, d\mu_g \\
&= \frac{1}{2} \int_\Sigma \langle \Delta^\perp \vec{\mathbf{H}} + g^{ik} g^{jl} \langle A_{ij}^0, \vec{\mathbf{H}} \rangle A_{kl}^0, V \rangle \, d\mu_g .
\end{aligned} \tag{1.3.6}$$

We abbreviate for a normal field ϕ

$$Q(A^0)\phi := g^{ik} g^{jl} \langle A_{ij}^0, \phi \rangle A_{kl}^0 \tag{1.3.7}$$

and

$$\delta \mathcal{W}(f) := \frac{1}{2} \Big(\Delta^\perp \vec{\mathbf{H}} + Q(A^0) \vec{\mathbf{H}} \Big). \tag{1.3.8}$$

Now we consider a general smooth one-parameter family of immersions $f_t : \Sigma \to \mathbb{R}^n$ with $\partial_t f_t = V_t \in \mathbb{R}^n$. We decompose

$$V_t = N_t + df.\xi_t$$

in N_t normal along f and $df.\xi_t$ tangential along f and $\xi_t \in T\Sigma$. We solve the ordinary differential equation

$$\phi_0 = id_\Sigma \quad \text{and} \quad \partial_t \phi_t = -\xi_t(\phi_t).$$

Clearly, $\phi_t : \Sigma \xrightarrow{\approx} \Sigma$ is a one-parameter family of diffeomorphisms. We put $\tilde{f}_t = f_t \circ \phi_t$ and see

$$\partial_t \tilde{f}_t = (\partial_t f_t) \circ \phi_t + df_t.\partial\phi_t = V_t(\phi_t) - df_t.\xi_t(\phi_t) = N_t(\phi_t)$$

which is normal along f. By parameter invariance of the Willmore functional and (1.3.6), we get

$$\frac{d}{dt}\mathcal{W}(f_t) = \frac{d}{dt}\mathcal{W}(\tilde{f}_t) = \int_\Sigma \langle \delta\mathcal{W}(f), N\rangle \, \mathrm{d}\mu_g = \int_\Sigma \langle \delta\mathcal{W}(f), \partial_t f\rangle \, \mathrm{d}\mu_g,$$

since $\delta\mathcal{W}(f)$ is normal along f. We have proved the following proposition.

Proposition 1.3.1. *For a smooth one-parameter family of immersions $f_t : \Sigma \to \mathbb{R}^n$ the first variation of the Willmore functional is given by*

$$\frac{d}{dt}\mathcal{W}(f_t) = \int_\Sigma \langle \delta\mathcal{W}(f), \partial_t f\rangle \, \mathrm{d}\mu_g. \tag{1.3.9}$$

f is called a Willmore immersion, if this vanishes, hence if

$$\delta\mathcal{W}(f) = \frac{1}{2}\Big(\Delta^\perp \vec{\mathbf{H}} + Q(A^0)\vec{\mathbf{H}}\Big) = 0 \quad on\ \Sigma. \tag{1.3.10}$$

In case of an immersion $f : \Sigma \to S^n$ into a sphere, we obtain the following proposition.

Proposition 1.3.2. *For an immersion $f : \Sigma \to S^n \subseteq \mathbb{R}^{n+1}$ into the sphere, we see for the second fundamental forms $A_{f,\mathbb{R}^{n+1}}$ respectively A_{f,S^n} as immersions into $\mathbb{R}^{n+1}$ respectively S^n*

$$\begin{gathered} A_{f,\mathbb{R}^{n+1}} = A_{f,S^n} - fg, \quad \vec{\mathbf{H}}_{f,\mathbb{R}^{n+1}} = \vec{\mathbf{H}}_{f,S^n} - 2f, \quad A^0_{f,\mathbb{R}^{n+1}} = A^0_{f,S^n}, \\ \Delta_g^{\mathbb{R}^{n+1},\perp}\vec{\mathbf{H}}_{f,\mathbb{R}^{n+1}} + Q(A^0_{f,\mathbb{R}^{n+1}})\vec{\mathbf{H}}_{f,\mathbb{R}^{n+1}} = \Delta_g^{S^n,\perp}\vec{\mathbf{H}}_{f,S^n} + Q(A^0_{f,S^n})\vec{\mathbf{H}}_{f,S^n}. \end{gathered} \tag{1.3.11}$$

Proof. Clearly, the second fundamental form A_{f,S^n} as map into the sphere is the orthogonal projection onto the tangent space of the sphere of the second fundamental form $A_{f,\mathbb{R}^{n+1}}$ as map into the Euclidean space, that is for the orthogonal projection $\pi_{S^n} : \mathbb{R}^{n+1} \to TS^n = \{f\}^\perp$, we have

$$A_{f,S^n} = \pi_{S^n} A_{f,\mathbb{R}^{n+1}} = A_{f,\mathbb{R}^{n+1}} - \langle A_{f,\mathbb{R}^{n+1}}, f\rangle f.$$

On the other hand, from $|f| \equiv 1$, we get $\partial f \perp f$ and

$$\langle A_{f,\mathbb{R}^{n+1},ij}, f\rangle = \langle \partial_{ij} f, f\rangle = -\langle \partial_i f, \partial_j f\rangle = -g_{ij},$$

hence

$$A_{f,\mathbb{R}^{n+1}} = A_{f,S^n} - fg,$$

and in particular $\vec{\mathbf{H}}_{f,\mathbb{R}^{n+1}} = \vec{\mathbf{H}}_{f,S^n} - 2f$ and $A^0_{f,\mathbb{R}^{n+1}} = A^0_{f,S^n}$.

Further $\Delta_g^{\mathbb{R}^{n+1},\perp} f = 0$ for the normal Laplacian, since ∇f is tangential. Observing for $\phi \in N_{S^n} f \subseteq TS^n \perp f$ that $\langle \partial\phi, f\rangle = -\langle \phi, \partial f\rangle = 0$, we see

$$\nabla^{\mathbb{R}^{n+1},\perp}\phi = \nabla^{S^n,\perp}\phi \perp f.$$

Therefore

$$\langle \nabla_i^{\mathbb{R}^{n+1},\perp}\nabla_j^{\mathbb{R}^{n+1},\perp}\phi, f\rangle = \langle \nabla_i \nabla_j^{S^n,\perp}\phi, f\rangle = -\langle \nabla_j^{S^n,\perp}\phi, \nabla_i f\rangle = 0,$$

hence

$$\nabla_i^{\mathbb{R}^{n+1},\perp}\nabla_j^{\mathbb{R}^{n+1},\perp}\phi = \nabla_i^{S^n,\perp}\nabla_j^{S^n,\perp}\phi,$$

in particular $\Delta_g^{\mathbb{R}^{n+1},\perp}\phi = \Delta_g^{S^n,\perp}\phi$. Together we obtain

$$\Delta_g^{\mathbb{R}^{n+1},\perp}\vec{\mathbf{H}}_{f,\mathbb{R}^{n+1}} + Q(A^0_{f,\mathbb{R}^{n+1}})\vec{\mathbf{H}}_{f,\mathbb{R}^{n+1}} = \Delta_g^{S^n,\perp}\vec{\mathbf{H}}_{f,S^n} + Q(A^0_{f,S^n})\vec{\mathbf{H}}_{f,S^n}$$

and (1.3.11) is proved. □

We see from (1.3.11) that

$$|\vec{\mathbf{H}}_{f,\mathbb{R}^{n+1}}|^2 = |\vec{\mathbf{H}}_{f,S^n}|^2 + 4.$$

Now for an immersion $f : \Sigma \to S^n$, we see by definition in (1.2.8) that

$$\begin{aligned} \mathcal{W}(f, S^n) &= \int_\Sigma \left(\frac{1}{4}|\vec{\mathbf{H}}_{f,S^n}|^2 + 1\right) \mathrm{d}\mu_f \\ &= \frac{1}{4}\int_\Sigma |\vec{\mathbf{H}}_{f,\mathbb{R}^{n+1}}|^2 \,\mathrm{d}\mu_f = \mathcal{W}(f, \mathbb{R}^{n+1}). \end{aligned} \tag{1.3.12}$$

f is a Willmore immersion in S^n, if and only if for any smooth one-parameter family of immersions $f_t : \Sigma \to S^n$

$$\frac{d}{dt}\mathcal{W}(f_t) = 0.$$

Observing $\partial_t f \in TS^n \perp f$ and using (1.3.9) and (1.3.11), we see

$$\begin{aligned} 0 = \frac{d}{dt}\mathcal{W}(f_t) &= \int_\Sigma \langle \delta\mathcal{W}(f), \partial_t f\rangle \, d\mu_g \\ &= \frac{1}{2}\int_\Sigma \langle \Delta_g^{\mathbb{R}^{n+1},\perp}\vec{\mathbf{H}}_{f,\mathbb{R}^{n+1}} + Q(A^0_{f,\mathbb{R}^{n+1}})\vec{\mathbf{H}}_{f,\mathbb{R}^{n+1}}, \partial_t f\rangle \, d\mu_g \\ &= \frac{1}{2}\int_\Sigma \langle \Delta_g^{S^n,\perp}\vec{\mathbf{H}}_{f,S^n} + Q(A^0_{f,S^n})\vec{\mathbf{H}}_{f,S^n}, \partial_t f\rangle \, d\mu_g, \end{aligned}$$

that is f is a Willmore immersion in the sphere if and only if

$$\begin{aligned} &\Delta_g^{S^n,\perp}\vec{\mathbf{H}}_{f,S^n} + Q(A^0_{f,S^n})\vec{\mathbf{H}}_{f,S^n} \\ &= \Delta_g^{\mathbb{R}^{n+1},\perp}\vec{\mathbf{H}}_{f,\mathbb{R}^{n+1}} + Q(A^0_{f,\mathbb{R}^{n+1}})\vec{\mathbf{H}}_{f,\mathbb{R}^{n+1}} \\ &= 0 \quad \text{on } \Sigma. \end{aligned} \tag{1.3.13}$$

In particular, f is a Willmore immersion in S^n, if and only if it is a Willmore immersion in $\mathbb{R}^{n+1}$.

By (1.3.10) minimal surfaces, that is when $\vec{\mathbf{H}} = 0$ are Willmore surfaces. By maximum principle, there are no closed minimal surfaces in $\mathbb{R}^n$. On the other hand by (1.3.13), minimal surfaces in S^n are Willmore surfaces. Let us consider the Clifford torus

$$T_{\text{Cliff}} := \frac{1}{\sqrt{2}}(S^1 \times S^1) \subseteq S^3$$

is a minimal surface in S^3, hence is a Willmore surface. We see from (1.2.8) that

$$\mathcal{W}(T_{\text{Cliff}}) = \text{Area}(T_{\text{Cliff}}) = \frac{1}{2}\mathcal{H}^1(S^1)^2 = 2\pi^2. \tag{1.3.14}$$

By conformal invariance of the Willmore functional, the stereographical image of T_{Cliff} is a Willmore surface in $\mathbb{R}^3$. Clearly, the stereographical projection $\Phi : S^3 \to \mathbb{R}^3 \cup \{\infty\}$ is given by

$$\Phi(x_1, x_2, x_3, x_4) := \frac{(x_1, x_2, x_3)}{1 - x_4}.$$

Now for $x = (\cos\alpha, \sin\alpha, \cos\beta, \sin\beta)/\sqrt{2} \in T_{\text{Cliff}}$, we get

$$\Phi(x) = \frac{(\cos\alpha, \sin\alpha, \cos\beta)}{\sqrt{2} - \sin\beta}.$$

We calculate

$$\begin{aligned}
|\Phi(x) - \sqrt{2}(\cos\alpha, \sin\alpha, 0)|^2 &= \left|\sqrt{2} - \frac{1}{\sqrt{2} - \sin\beta}\right|^2 + \frac{\cos^2\beta}{(\sqrt{2} - \sin\beta)^2} \\
&= \frac{|\sqrt{2}(\sqrt{2} - \sin\beta) - 1|^2 + \cos^2\beta}{(\sqrt{2} - \sin\beta)^2} \\
&= \frac{|1 - \sqrt{2}\sin\beta|^2 + \cos^2\beta}{(\sqrt{2} - \sin\beta)^2} \\
&= \frac{1 - 2\sqrt{2}\sin\beta + 2\sin^2\beta + \cos^2\beta}{2 - 2\sqrt{2}\sin\beta + \sin^2\beta} = 1.
\end{aligned}$$

Abbreviating $e_\alpha = (\cos\alpha, \sin\alpha, 0)$, we see

$$\Phi(x) \in \partial B_1(\sqrt{2}e_\alpha) \cap \text{span}\{e_\alpha, e_3\} = \{(\sqrt{2} + \cos\gamma)e_\alpha + \sin\gamma\ e_3\}$$

and

$$\Phi(T_{\text{Cliff}}) \subseteq \{(\sqrt{2} + \cos\gamma)(\cos\alpha, \sin\alpha, 0) + \sin\gamma\ e_3\} =: T_{\text{round}}.$$

T_{round} is a rotational round torus with radi $\sqrt{2}$ and 1. Since T_{round} is a closed surface and

$$\Phi : T_{\text{Cliff}} \to \{(\sqrt{2} + \cos\gamma)(\cos\alpha, \sin\alpha, 0) + \sin\gamma e_3\} =: T_{\text{round}}$$

is a local diffeomorphism, $\Phi(T_{\text{Cliff}}) \subseteq T_{\text{round}}$ is open. On the other hand, $\Phi(T_{\text{Cliff}})$ is compact, as T_{Cliff} is compact, hence $\Phi(T_{\text{Cliff}}) = T_{\text{round}}$, and T_{round} is the stereographical image of the Clifford torus, hence a Willmore surface.

Willmore conjectured that the Clifford torus or more precisely T_{round} is the unique minimizer among all tori in $\mathbb{R}^3$, see [Wil65], in particular recalling (1.3.14)

$$\mathcal{W}(T^2) \geq 2\pi^2 \tag{1.3.15}$$

for any torus $T^2 \subseteq \mathbb{R}^3$.

The existence of a minimizing torus in $\mathbb{R}^n$ was proved in [Sim93], and when combined with [BaKu03], this gives also the existence of a minimizer for closed, orientable surfaces of any genus.

References

[BaKu03] M. BAUER and E. KUWERT, *Existence of Minimizing Willmore Surfaces of Prescribed Genus*, IMRN Intern. Math. Res. Notes **10** (2003), 553–576.

[Ch74] B. Y. CHEN, *Some conformal Invariants of Submanifolds and their Application*, Bollettino della Unione Matematica Italiana, Serie 4 **10** (1974) 380–385.

[dC] M. P. DO CARMO, “Riemannian Geometry”, Birkhäuser, Boston - Basel - Berlin, 1992.

[Sim93] L. SIMON, *Existence of surfaces minimizing the Willmore functional*, Communications in Analysis and Geometry, **1**, No. 2 (1993), 281–326.

[Wei78] J. WEINER, *On a problem of Chen, Willmore, et al.*, Indiana University Mathematical Journal **27** (1978), 18–35.

[Wil65] T. J. WILLMORE, *Note on Embedded Surfaces*, An. Stiint. Univ. “Al. I. Cusa” Iasi Sect. Ia, **11B** (1965), 493–496.

[Wil82] T. J. WILLMORE, “Total curvature in Riemannian Geometry”, Wiley, 1982.

2 Monotonicity formula

2.1 Monotonicity formula

In this section, we review the arguments in [Sim93], see also [KuSch04], proving a monotonicity formula for properly immersed, open surfaces $f : \Sigma \to U \subseteq \mathbb{R}^n$ open, with square integrable mean curvature

$$\mathcal{W}(f) = \int_\Sigma |\vec{\mathbf{H}}_f|^2 \, d\mu_f < \infty.$$

As f is a proper immersion into U, we see that $\mu = f(\mu_f)$ is a Radon measure in U.

The divergence theorem for manifolds gives

$$\int_\Sigma \operatorname{div}_\Sigma \eta \, d\mu = - \int_\Sigma \eta \vec{\mathbf{H}} \, d\mu \quad \forall \eta \in C_0^1(U, \mathbb{R}^n). \tag{2.1.1}$$

Approximating this holds true for Lipschitz test functions η. We consider $B_\varrho(x_0) \Subset U, 0 < \sigma < \varrho$ and put $|x|_\sigma := \max(|x|, \sigma)$ and $\eta(x) := (|x - x_0|_\sigma^{-2} - \varrho^{-2})_+ \cdot (x - x_0)$. Observing $\operatorname{div}_\Sigma x = 2$ and decomposing $(x - x_0) = (x - x_0)^{\tan} + (x - x_0)^\perp$ in tangential and normal components with respect to Σ, we calculate

$$\begin{aligned}
\operatorname{div}_\Sigma \eta(x) &= 2(|x - x_0|_\sigma^{-2} - \varrho^{-2})_+ - 2\frac{(x - x_0)^{\tan}}{|x - x_0|^4}(x - x_0)\chi_{B_\varrho(x_0) - B_\sigma(x_0)} \\
&= 2\sigma^{-2}\chi_{B_\sigma(x_0)} - 2\varrho^{-2}\chi_{B_\varrho(x_0)} + 2\frac{|(x - x_0)^\perp|^2}{|x - x_0|^4}\chi_{B_\varrho(x_0) - B_\sigma(x_0)}
\end{aligned}$$

and obtain from (2.1.1) that

$$\begin{aligned}
&2\sigma^{-2}\mu(B_\sigma(x_0)) + 2 \int_{\Sigma \cap B_\varrho(x_0) - B_\sigma(x_0)} \frac{|(x - x_0)^\perp|^2}{|x - x_0|^4} \, d\mu \\
&= 2\varrho^{-2}\mu(B_\varrho(x_0)) - \int_{\Sigma \cap B_\varrho(x_0)} (|x - x_0|_\sigma^{-2} - \varrho^{-2})_+ \cdot (x - x_0)\vec{\mathbf{H}} \, d\mu.
\end{aligned}$$

By using the identity

$$\frac{|(x - x_0)^\perp|^2}{|x - x_0|^4} + \frac{1}{2}\frac{x - x_0}{|x - x_0|^2}\vec{\mathbf{H}} = \left|\frac{1}{4}\vec{\mathbf{H}} + \frac{(x - x_0)^\perp}{|x - x_0|^2}\right|^2 - \frac{1}{16}|\vec{\mathbf{H}}|^2$$

we then calculate

$$\begin{aligned}
&\sigma^{-2}\mu(B_\sigma(x_0)) + \int\limits_{B_\varrho(x_0)-B_\sigma(x_0)} \left|\frac{1}{4}\vec{\mathbf{H}} + \frac{(x-x_0)^\perp}{|x-x_0|^2}\right|^2 \, d\mu \\
&= \varrho^{-2}\mu(B_\varrho(x_0)) + \frac{1}{16} \int\limits_{B_\varrho(x_0)-B_\sigma(x_0)} |\vec{\mathbf{H}}|^2 \, d\mu \\
&\quad + \frac{1}{2} \int\limits_{B_\varrho(x_0)} \varrho^{-2}(x-x_0)\, \vec{\mathbf{H}}(x) \, d\mu(x) \\
&\quad - \frac{1}{2} \int\limits_{B_\sigma(x_0)} \sigma^{-2}(x-x_0)\, \vec{\mathbf{H}}(x) \, d\mu(x).
\end{aligned} \tag{2.1.2}$$

Putting

$$R_{x_0,\varrho} := \frac{1}{2} \int\limits_{B_\varrho(x_0)} \varrho^{-2}(x-x_0)\, \vec{\mathbf{H}}(x) \, d\mu(x)$$

and

$$\gamma(\varrho) := \varrho^{-2}\mu(B_\varrho(x_0)) + \frac{1}{16} \int\limits_{B_\varrho(x_0)} |\vec{\mathbf{H}}|^2 \, d\mu + R_{x_0,\varrho}, \tag{2.1.3}$$

we see that γ is monotonically nondecreasing. We estimate

$$|R_{x_0,\varrho}| \le \frac{1}{2}\Big(\varrho^{-2}\mu(B_\varrho(x_0))\Big)^{1/2} \, \| \vec{\mathbf{H}} \|_{L^2(B_\varrho(x_0))} \tag{2.1.4}$$

and get for any $\delta > 0$

$$\begin{aligned}
\sigma^{-2}\mu(B_\sigma(x_0)) &\le \varrho^{-2}\mu(B_\varrho(x_0)) + \frac{1}{16} \int\limits_{B_\varrho(x_0)} |\vec{\mathbf{H}}|^2 \, d\mu \\
&\quad + \frac{1}{2}\Big(\varrho^{-2}\mu(B_\varrho(x_0))\Big)^{1/2} \, \| \vec{\mathbf{H}} \|_{L^2(B_\varrho(x_0))} \\
&\quad + \frac{1}{2}\Big(\sigma^{-2}\mu(B_\sigma(x_0))\Big)^{1/2} \, \| \vec{\mathbf{H}} \|_{L^2(B_\sigma(x_0))} \\
&\le (1+\delta)\varrho^{-2}\mu(B_\varrho(x_0)) \\
&\quad + \left(\frac{1}{16} + C\delta^{-1}\right) \int\limits_{B_\varrho(x_0)} |\vec{\mathbf{H}}|^2 \, d\mu + \delta\sigma^{-2}\mu(B_\sigma(x_0)),
\end{aligned}$$

in particular for $0 < \varrho \le \varrho_0$

$$\varrho^{-2}\mu(B_\varrho(x_0)) \le (1+\delta)\varrho_0^{-2}\mu(B_{\varrho_0}(x_0)) + C(1+\delta^{-1})\mathcal{W}(f) < \infty. \tag{2.1.5}$$

Then we get from (2.1.4) that

$$\lim_{\varrho \downarrow 0} R_{x_0,\varrho} = 0, \tag{2.1.6}$$

hence the density

$$\theta^2(\mu, x_0) \text{ exists} \tag{2.1.7}$$

and

$$\omega_2 \theta^2(\mu, x_0) \le \varrho^{-2}\mu(B_\varrho(x_0)) + \frac{1}{16}\int\limits_{B_\varrho(x_0)} |\vec{\mathbf{H}}|^2 \,\mathrm{d}\mu + R_{x_0,\varrho}. \tag{2.1.8}$$

For $x_j \to x_0, 0 < \varrho < \varrho_0/2$, we get

$$\begin{aligned}
\varrho^{-2}\mu(\overline{B_\varrho(x_0)}) &\ge \limsup_{j\to\infty} \varrho^{-2}\mu(B_\varrho(x_j)) \\
&\ge \limsup_{j\to\infty} \left(\omega_2\theta^2(\mu, x_j) - \frac{1}{16}\int\limits_{B_\varrho(x_j)} |\vec{\mathbf{H}}|^2 \,\mathrm{d}\mu - R_{x_j,\varrho} \right) \\
&\ge \limsup_{j\to\infty} \ \omega_2\theta^2(\mu, x_j) \\
&\quad - C\Big(\varrho_0^{-2}\mu(B_{\varrho_0}(x_0)) + \mathcal{W}(f)\Big)^{1/2} \parallel \vec{\mathbf{H}} \parallel_{L^2(B_{2\varrho}(x_0))} .
\end{aligned}$$

Letting $\varrho \downarrow 0$ this yields

$$\theta^2(\mu, x_0) \ge \limsup_{j\to\infty} \theta^2(\mu, x_j), \tag{2.1.9}$$

and $\theta^2(\mu)$ is upper semicontinuous.

Now we consider $U = \mathbb{R}^n$. If $\limsup_{\varrho\to\infty} \varrho^{-2}\mu(B_\varrho(0)) = \infty$ then by (2.1.5)

$$\lim_{\varrho\to\infty} \varrho^{-2}\mu(B_\varrho(0)) = \infty \tag{2.1.10}$$

and by (2.1.4)

$$\lim_{\varrho\to\infty} \gamma(\varrho) = \infty. \tag{2.1.11}$$

If $\limsup_{\varrho\to\infty} \varrho^{-2}\mu(B_\varrho(0)) < \infty$, we estimate for $0 < \sigma < \varrho < \infty$

$$|R_{0,\varrho}| \leq \frac{1}{2\varrho} \int\limits_{B_\varrho(0)} |\vec{\mathbf{H}}| \, d\mu$$
$$\leq \frac{1}{2\varrho} \int\limits_{B_\sigma(0)} |\vec{\mathbf{H}}| \, d\mu + \frac{1}{2}\Big(\varrho^{-2}\mu(B_\varrho(0))\Big)^{1/2} \parallel \vec{\mathbf{H}} \parallel_{L^2(\mathbb{R}^n - B_\sigma(0))},$$

hence

$$\limsup_{\varrho\to\infty} |R_{0,\varrho}| \leq \frac{1}{2}\Big(\limsup_{\varrho\to\infty} \varrho^{-2}\mu(B_\varrho(0))\Big)^{1/2} \parallel \vec{\mathbf{H}} \parallel_{L^2(\mathbb{R}^n - B_\sigma(0))}$$

and letting $\sigma \to \infty$

$$\lim_{\varrho\to\infty} R_{x_0,\varrho} = 0. \tag{2.1.12}$$

As γ is monotonically nondecreasing, we see in any case by (2.1.10), (2.1.11) and (2.1.12) that

$$\lim_{\varrho\to\infty} \gamma(\varrho) = \lim_{\varrho\to\infty} \varrho^{-2}\mu(B_\varrho(0)) + \frac{1}{4}\mathcal{W}(f) \in [0,\infty] \quad \text{exists}, \tag{2.1.13}$$

in particular the density at infinity

$$\theta^2(\mu,\infty) := \lim_{\varrho\to\infty} \frac{\mu(B_\varrho(0))}{\omega_2\varrho^2} \in [0,\infty]$$

exists. Letting $\varrho \to \infty$ in (2.1.8) and recalling (2.1.12) and $\omega_2 = \pi$, we get

$$\theta^2(\mu,.) \leq \theta^2(\mu,\infty) + \frac{1}{4\pi}\mathcal{W}(f). \tag{2.1.14}$$

If $\theta^2(\mu,\infty) = 0$, which is certainly true when $\operatorname{spt}\mu$ is compact, we get from (2.1.5)

$$\varrho^{-2}\mu(B_\varrho(x_0)) \leq C\mathcal{W}(f). \tag{2.1.15}$$

Moreover observing for $x_0 \in \operatorname{spt}\mu$ that

$$\theta^2(\mu,x_0) = \#(f^{-1}(x_0)) \geq 1, \tag{2.1.16}$$

we arrive in the following proposition at the Li-Yau-inequality, see [LY82].

Proposition 2.1.1. *For any immersion* $f : \Sigma \to \mathbb{R}^n$ *of a closed surface and the image* $\mu = f(\mu_f)$ *of the induced area measure, we have*

$$\#(f^{-1}(.)) = \theta^2(\mu, .) \leq \frac{1}{4\pi}\mathcal{W}(f), \tag{2.1.17}$$

in particular

$$\mathcal{W}(f) \geq 4\pi, \tag{2.1.18}$$

and if

$$\mathcal{W}(f) < 8\pi \tag{2.1.19}$$

then f *is an embedding.*

Moreover equality in (2.1.18) *implies that* f *parametrizes a round sphere.*

Proof. The three inequalities follow directly from $\theta^2(\mu, \infty) = 0$ and (2.1.14).

In case of equality in (2.1.18), we may assume by (2.1.19) that f is an embedding, and we consider a closed surface $\Sigma \subseteq \mathbb{R}^n$ with $\mathcal{W}(\Sigma) = 4\pi$. We choose $x_0 \in \Sigma \neq \emptyset$ and see from monotonicity of γ in (2.1.3) using (2.1.6) and (2.1.12) that

$$\begin{aligned} \pi = \#(f^{-1}(x_0))\pi &= \lim_{\varrho \to 0} \gamma(\varrho) \leq \gamma(r) \leq \lim_{\varrho \to \infty} \gamma(\varrho) \\ &= \frac{1}{4}\mathcal{W}(\Sigma) = \pi \quad \forall\, r, \end{aligned}$$

hence $\gamma \equiv \pi$ is constant. Then by (2.1.2)

$$\vec{\mathbf{H}}(x) + 4\frac{(x - x_0)^\perp}{|x - x_0|^2} = 0 \quad \forall x \neq x_0 \in \Sigma,$$

where $^\perp$ denotes the orthogonal projection onto $N_x\Sigma$. Now by maximum principle, we know that Σ is not minimal. Hence by translating, we may assume that $x = 0 \in \Sigma$ with $\vec{\mathbf{H}}(x) \neq 0$. After rotating and rescaling, we may further assume $T_0\Sigma = \operatorname{span}\{e_1, e_2\}$, $N_0\Sigma = \operatorname{span}\{e_3, \ldots, e_n\}$, $\vec{\mathbf{H}}(0) = 2e_3$. Then replacing x_0 by x, we firstly get from above for $j = 4, \ldots, n$, that

$$0 = \langle \vec{\mathbf{H}}(0), e_j \rangle = -4\left\langle \frac{-x^\perp}{|x|^2}, e_j \right\rangle = 4\left\langle \frac{x^\perp}{|x|^2}, e_j \right\rangle = 4x_j/|x|^2$$
$$\text{for all } x \neq 0 \in \Sigma,$$

hence $\Sigma \subseteq \mathbb{R}^3$. Next for $j = 3$

$$2 = \langle \vec{\mathbf{H}}(0), e_3 \rangle = -4\left\langle \frac{-x^\perp}{|x|^2}, e_3 \right\rangle = 4x_3/|x|^2 \quad \text{for all } x \neq 0 \in \Sigma,$$

hence $2x_3 = |x|^2$ or likewise

$$1 = x_1^2 + x_2^2 + x_3^2 - 2x_3 + 1 = x_1^2 + x_2^2 + (1 - x_3)^2 = |x - e_3|^2$$

and $\Sigma \subseteq \partial_1 B(e_3)$. Together we see that $\Sigma \subseteq \partial B_1(e_3) \cap \mathbb{R}^3 \cong S^2$, hence $\Sigma = \partial B_1(e_3)$, as Σ is compact and $\partial\Sigma = \emptyset$, and the proposition is proved. □

Proposition 2.1.2. *For any proper immersion $f : \Sigma \to B_\varrho(x_0)$ of an open, connected surface Σ and the image $\mu = f(\mu_f)$ of the induced area measure, we have for $x_0 \in f(\Sigma)$, $\overline{f(\Sigma)} \not\subseteq B_\varrho(x_0)$, that*

$$\mu(B_\varrho(x_0)) \geq c_0 \varrho^2 / \mathcal{W}(f) \tag{2.1.20}$$

for some $c_0 > 0$.

Proof. We select an integer N and see, as $x_0 \in f(\Sigma)$, $\overline{f(\Sigma)} \not\subseteq B_\varrho(x_0)$ and $f(\Sigma)$ is connected, that there exist

$$x_j \in f(\Sigma) \cap \partial B_{(j+1/2)\varrho/N}(x_0) \neq \emptyset \quad \text{for } j = 1, \ldots, N-1.$$

By (2.1.5) and (2.1.16)

$$1 \leq \theta^2(\mu, x_j) \leq C(\varrho/(2N))^{-2} \mu(B_{\varrho/(2N)}(x_j)) + C \parallel \vec{\mathbf{H}}_\mu \parallel^2_{L^2(B_{\varrho/(2N)}(x_j))}$$

and adding up observing that $B_{\varrho/(2N)}(x_j)$, $j = 0, \ldots, N-1$ are pairwise disjoint,

$$N \leq C N^2 \varrho^{-2} \mu(B_\varrho(x_0)) + C\mathcal{W}(f).$$

If $\mathcal{W}(f)/(\varrho^{-2}\mu(B_\varrho(x_0))) \leq 1$, we have (2.1.20) with $c_0 = 1$. Otherwise we can select an integer $N \leq \sqrt{\mathcal{W}(f)/(\varrho^{-2}\mu(B_\varrho(x_0)))} \leq N + 1 \leq 2N$. This yields

$$1 \leq C\sqrt{\mathcal{W}(f)\varrho^{-2}\mu(B_\varrho(x_0))}$$

which implies (2.1.20). □

Proposition 2.1.3. *For any immersion $f : \Sigma \to \mathbb{R}^n$ of a closed surface and the image $\mu = f(\mu_f)$ of the induced area measure, we have*

$$\mu_f(\Sigma)/\mathcal{W}(f) \leq \operatorname{diam} f(\Sigma)^2 \leq C\mu_f(\Sigma)\,\mathcal{W}(f). \tag{2.1.21}$$

Proof. Plugging $\eta(y) := y - x$ with $x \in f(\Sigma)$ into (2.1.1), we see $\operatorname{div}_\Sigma \eta = 2$ on Σ and

$$2\mu_f(\Sigma) = -\int_\Sigma (y - x)\vec{\mathbf{H}}(y)\, d\mu(y) \leq \operatorname{diam} f(\Sigma) \int_\Sigma |\vec{\mathbf{H}}_f|\, d\mu$$

$$\leq \operatorname{diam} f(\Sigma)(4\mathcal{W}(f))^{1/2}\mu_f(\Sigma)^{1/2},$$

which is the inequality on the left in (2.1.21).

Choosing $x_0, y_0 \in f(\Sigma)$ with $|x_0 - y_0| = \operatorname{diam} f(\Sigma) =: d$, we see $x_0 \in f(\Sigma) \not\subseteq B_d(x_0)$ and by (2.1.20)

$$c_0 d^2/\mathcal{W}(f) \leq \mu(B_d(x_0)) \leq \mu_f(\Sigma),$$

which is the inequality on the right in (2.1.21). □

Proposition 2.1.4. *For any proper immersion $f : \Sigma \to \mathbb{R}^n$ of an open surface Σ and the image $\mu = f(\mu_f)$ of the induced area measure, we have*

$$\#(\textit{components of}\operatorname{spt}\mu) \leq \theta^2(\mu, \infty) + \frac{1}{4\pi}\mathcal{W}(f). \tag{2.1.22}$$

In particular, if $\theta^2(\mu, \infty) < \infty$, then $\operatorname{spt}\mu$ has only finitely many components.

Proof. If $\operatorname{spt}\mu$ consists of at least N components, we can find in successive stages open pairwise disjoint subsets $U_1, \ldots, U_N \subseteq \mathbb{R}^n$ with

$$\operatorname{spt}\mu \subseteq U_1 \cup \ldots \cup U_N,$$
$$U_i \cap \operatorname{spt}\mu \neq \emptyset \quad \text{for all } i = 1, \ldots, N,$$

hence by (2.1.14) and (2.1.16)

$$N \leq \sum_{i=1}^{N} \left(\theta^2(\mu\lfloor U_i, \infty) + \frac{1}{4\pi}\mathcal{W}(f\lfloor f^{-1}(U_i))\right) = \theta^2(\mu, \infty) + \frac{1}{4\pi}\mathcal{W}(f),$$

and (2.1.22) follows. □

Actually, the assumption $\theta^2(\mu, \infty) = 0$ is equivalent to the compactness of the support of μ. More precisely, we have the following proposition.

Proposition 2.1.5. *For any proper immersion $f : \Sigma \to \mathbb{R}^n$ of an open surface Σ and the image $\mu = f(\mu_f)$ of the induced area measure, we have*

$$\theta^2(\mu, \infty) \geq 1 \Longleftrightarrow \operatorname{spt}\mu \textit{ is not compact}. \tag{2.1.23}$$

Proof. The inclusion from left to the right is obvious.

Now we assume that $\operatorname{spt}\mu$ is not compact. Then $\operatorname{spt}\mu$ is unbounded, and there exist $x_j \in \operatorname{spt}\mu$ with $2\varrho_j := |x_j| \to \infty$. From (2.1.5) and (2.1.16), we get for any $\delta > 0$

$$\pi \leq (1+\delta)\varrho_j^{-2}\mu(B_{\varrho_j}(x_j)) + C_\delta \int\limits_{B_{\varrho_j}(x_j)} |\vec{\mathbf{H}}|^2 \, d\mu$$

$$\leq 9(1+\delta)(3\varrho_j)^{-2}\mu(B_{3\varrho_j}(0)) + C_\delta \int\limits_{\mathbb{R}^n - B_{\varrho_j}(0)} |\vec{\mathbf{H}}|^2 \, d\mu.$$

Letting first $j \to \infty$ and then $\delta \downarrow 0$, we get

$$\theta^2(\mu, \infty) = \lim_{j\to\infty} \frac{\mu(B_{3\varrho_j}(0))}{\pi(3\varrho_j)^2} \geq 1/9,$$

which is (2.1.23) with 1 replaced by $1/9$.

To prove the full strength of (2.1.23), we may assume that $\theta^2(\mu, \infty) < \infty$. Then the limit

$$\zeta_{\varrho_j,\#}\mu \to \nu \quad \text{exists weakly as Radon measures}$$

for some sequence $\varrho_j \to \infty$. We see for any $\varrho > 0$ that

$$\frac{\nu(\overline{B_\varrho(0)})}{\omega_2 \varrho^2} \geq \liminf_{j\to\infty} \frac{\mu(B_{\varrho_j \varrho}(0))}{\omega_2(\varrho_j \varrho)^2} \geq 1/9$$

and $0 \in \operatorname{spt} \nu$. Further, ν is an stationary integral varifold by Allard's integral compactness theorem, see [All72, Theorem 6.4] or [Sim, Remark 42.8]. In particular $\vec{\mathbf{H}}_\nu = 0$ and starting with ν in (2.1.1), we get by upper semicontinuity of $\theta^2(\nu, .)$ and integrality of ν that $\theta^2(\nu, .) \geq 1$ on $\operatorname{spt} \nu$ and by (2.1.8)

$$1 \leq \theta^2(\nu, 0) \leq \frac{\nu(B_1(0))}{\omega_2} \leq \liminf_{j\to\infty} \frac{\mu(B_{\varrho_j}(0))}{\omega_2 \varrho_j^2} = \theta^2(\mu, \infty),$$

which establishes (2.1.23). □

Finally, we consider f_j with $\mathcal{W}(f_j) < \infty$ and $\mu_{f_j} \to \mu = \mu_f$ weakly as Radon measures. For γ_{f_j} defined in (2.1.3) by μ_{f_j}, we see

$$\gamma(\varrho) \leq \liminf_{j\to\infty} \gamma_{f_j}(\varrho) \quad \text{for almost all } \varrho > 0,$$

hence by (2.1.13), (2.1.14) and monotonicity of γ_{f_j}

$$\begin{aligned} \theta^2(\mu, \cdot) &\leq \theta^2(\mu, \infty) + \frac{1}{4\pi}\mathcal{W}(f) \\ &\leq \liminf_{j\to\infty} \Big(\theta^2(\mu_{f_j}, \infty) + \frac{1}{4\pi}\mathcal{W}(f_j)\Big). \end{aligned} \tag{2.1.24}$$

References

[All72] W. K. Allard, *On the first variation of a varifold*, Annals of Mathematics **95** (1972), 417–491.

[KuSch04] E. KUWERT and R. SCHÄTZLE, *Removability of point singularities of Willmore surfaces*, Annals of Mathematics **160** (2004) 315–357.

[LY82] P. LI and S. T. YAU, *A new conformal invariant and its applications to the Willmore conjecture and the first eigenvalue on compact surfaces*, Inventiones Mathematicae **69** (1982), 269–291.

[Sim] L. SIMON, "Lectures on Geometric Measure Theory", Proceedings of the Centre for Mathematical Analysis Australian National University, Vol. 3, 1983.

[Sim93] L. SIMON, *Existence of surfaces minimizing the Willmore functional*, Communications in Analysis and Geometry, **1** (1993), 281–326.

3 The Willmore flow

3.1 Introduction

The gradient flow for the Wilmore functional is called the Willmore flow. According to the formula (1.3.9) for the first variation, this is the evolution of a smooth family of immersions $f_t : \Sigma \to \mathbb{R}^n$ satisfying the parabolic equation $\partial_t^\perp f_t + \delta \mathcal{W}(f_t) = 0$ or by (1.3.10)

$$\partial_t^\perp f_t = -\frac{1}{2}\Big(\Delta_g^\perp \vec{\mathbf{H}} + Q(A^0)\vec{\mathbf{H}}\Big), \tag{3.1.1}$$

where $\partial_t^\perp$ denotes the normal projection of the time derivative as in Section 1.3. This is a quasi-linear fourth order parabolic equation. Due to parameter invariance, it is only degenerate parabolic, and existence of solutions for a short time is not immediate.

As in Section 1 we decompose

$$\partial_t f_t = N_t + df.\xi_t$$

in $N_t = -\delta \mathcal{W}(f_t)$ normal along f and $df.\xi_t$ tangential along f and $\xi_t \in T\Sigma$. We solve the ordinary differential equation

$$\phi_0 = id_\Sigma \quad \text{and} \quad \partial_t \phi_t = -\xi_t(\phi_t).$$

Clearly, $\phi_t : \Sigma \xrightarrow{\approx} \Sigma$ is a one-parameter family of diffeomorphisms. We put $\tilde{f}_t = f_t \circ \phi_t$ and see

$$\begin{aligned} \partial_t \tilde{f}_t &= (\partial_t f_t) \circ \phi_t + df_t.\partial\phi_t = \partial_t f_t(\phi_t) - df_t.\xi_t(\phi_t) = N_t(\phi_t) \\ &= -\delta \mathcal{W}(f_t) \circ \phi_t = -\delta \mathcal{W}(\tilde{f}_t) \end{aligned}$$

which is a normal Willmore flow.

On the other hand comming back to (3.1.1), we see for small times that

$$\| f_t - f_0 \|_{C^{m,\alpha}} < \varepsilon$$

for $\alpha, \varepsilon > 0$. For $m \geq 1$ and $\varepsilon > 0$ small enough, there exist diffeomorphisms $\phi_t : \Sigma \xrightarrow{\approx} \Sigma$ and vector fields $N_t : \Sigma \to \mathbb{R}^n$ which are normal along f_0 such that

$$f_t \circ \phi_t = f_0 + N_t =: \tilde{f}_t.$$

Moreover,

$$\| N_t \|_{C^{m,\alpha}} = \| \tilde{f}_t - f_0 \|_{C^{m,\alpha}} \leq C\varepsilon \tag{3.1.2}$$

for some constant $C < \infty$ independent of ε, if ε is small enough.

Clearly by parameter invariance (3.1.1) transforms into

$$\partial_t^\perp \tilde{f}_t = \partial_t^\perp f_t \circ \phi_t = -\delta\mathcal{W}(f_t) \circ \phi_t = -\delta\mathcal{W}(\tilde{f}_t). \tag{3.1.3}$$

We write $P^\perp_{\tilde{f}_t}$, $P^\perp_{f_0}$ for the orthognonal projections of $\mathbb{R}^n$ onto $N\tilde{f}_t$, Nf_0. Choosing locally a smooth basis $\nu_1, \ldots, \nu_{n-2}$ of $Nf_0 = P^\perp_{f_0}\mathbb{R}^n$, and writing $N_t = \varphi_{r,t}\nu_r$, we obtain from (3.1.3) that

$$\begin{aligned}
2(\partial_t\varphi_{r,t})\, P^\perp_{\tilde{f}_t}\nu_r &= -\Delta^\perp\vec{\mathbf{H}} - Q(A^0)\vec{\mathbf{H}} \\
&= -g^{ij} P^\perp_{\tilde{f}_t}\nabla_i P^\perp_{\tilde{f}_t}\nabla_j \Delta_g \tilde{f}_t + B(\cdot, N_t, DN_t, D^2N_t) \\
&= -g^{ij} g^{kl} P^\perp_{\tilde{f}_t}\nabla_i P^\perp_{\tilde{f}_t}\nabla_j (\partial_k\partial_l \tilde{f}_t - \Gamma^m_{kl}\partial_m \tilde{f}_t) \\
&\quad + B(\cdot, N_t, DN_t, D^2N_t) \qquad\qquad (3.1.4)\\
&= -g^{ij} g^{kl} P^\perp_{\tilde{f}_t}\partial_{ijkl}N_t + B(\cdot, N_t, DN_t, D^2N_t, D^3N_t) \\
&= -(g^{ij} g^{kl}\partial_{ijkl}\varphi_{r,t}) P^\perp_{\tilde{f}_t}\nu_r \\
&\quad + B(\cdot, N_t, DN_t, D^2N_t, D^3N_t),
\end{aligned}$$

where B is a smooth function depending on f_0 and changing from line to line. For $\|\tilde{f}_t - f_0\|_{C^1} \leq \delta$ small enough, $g^{ij}g^{kl}$ is uniformly elliptic, and there exists a constant $c_0 > 0$ such that

$$|P^\perp_{\tilde{f}_t}N| \geq |N| - |P_{f_0}N - P_{\tilde{f}_t}N| \geq c_0|N| \quad \text{for } N \text{ normal along } f_0, \tag{3.1.5}$$

hence $P^\perp_{\tilde{f}_t}\nu_1, \ldots, P^\perp_{\tilde{f}_t}\nu_{n-2}$ forms a basis of $N_{\tilde{f}_t}$. Therefore,

$$\partial_t\varphi_{r,t} + \frac{1}{2}g^{ij}\, g^{kl}\partial_{ijkl}\varphi_{r,t} = B_r(\cdot, \varphi_t, D\varphi_t, D^2\varphi_t, D^3\varphi_t), \tag{3.1.6}$$

and by (3.1.2)

$$\|\varphi_{r,t}\|_{C^{m,\alpha}} \leq C\varepsilon \quad \text{for } r = 1, \ldots, n-2. \tag{3.1.7}$$

Since the coefficients $g^{ij}g^{kl}$ are uniformly elliptic for $\|\varphi_{r,t}\|_{C^{m,\alpha}} \leq \delta$ small enough, (3.1.6) admits a unique local solution in the Hölder space $C^{m,\alpha}(\Sigma, \mathbb{R}^n)$ for $m \geq 4$. This follows from Schauder estimates for linear uniformly parabolic equations in Hölder spaces, see [Ei], and by means of fixed-point theorems. This argument shows that (3.1.1) has for given smooth initial data a solution locally in time which is unique up to reparametrization.

3.2 Estimation of the maximal existence time

In this section, we want to estimate the maximal existence time for a Willmore flow.

Theorem 3.2.1 ([KuSch02, Theorem 1.2]). *Let* $f_t : \Sigma \to \mathbb{R}^n$ *be a normal Willmore flow* $\partial_t f_t + \delta\mathcal{W}(f_t) = 0$ *on the maximal existence interval* $[0, T[$ *with* $0 < T \leq \infty$. *For appropriate* $\varepsilon_0 = \varepsilon_0(n) > 0$, *we consider* $\varrho > 0$ *satisfying*

$$\int_{B_\varrho(x)} |A_0|^2 \, \mathrm{d}\mu_{f_0} \leq \varepsilon \leq \varepsilon_0 \quad \textit{for all } x \in \mathbb{R}^n. \tag{3.2.1}$$

Then

$$T > c_0 \varrho^4 \tag{3.2.2}$$

and

$$\int_{B_\varrho(x)} |A_t|^2 \, \mathrm{d}\mu_{f_t} \leq C\varepsilon \quad \textit{for } 0 \leq t \leq c_0\varrho^4, x \in \mathbb{R}^n \tag{3.2.3}$$

for $0 < c_0, C < \infty$ *depending only on* n.

Sketch of the Proof. To abbreviate notation for normal tensors ϕ, ψ, we denote by $\phi * \psi$ any normal tensor obtained by contraction in a bilinear way, and we write

$$P_r^m(A) = \sum_{i_1+\ldots+i_r=m} \nabla^{\perp,i_1} A * \nabla^{\perp,i_2} A * \ldots * \nabla^{\perp,i_r} A.$$

Clearly,

$$P_r^m(A) * P_s^l(A) = P_{r+s}^{m+l}(A) \quad \text{and} \quad \nabla^\perp P_r^m(A) = P_r^{m+1}(A).$$

We recall for a normal evolution $\partial_t f = V \in Nf$ by (1.3.4) that

$$\partial_t^\perp A_{ij} = \nabla_i^\perp \nabla_j^\perp V - \langle A_{jk}, V\rangle g^{kl} A_{il} = \nabla_i^\perp \nabla_j^\perp V + A * A * V.$$

In case of a normal Willmore flow, we have

$$V = -\frac{1}{2}\left(\Delta^\perp \vec{\mathbf{H}} + Q(A^0)\vec{\mathbf{H}}\right) = P_1^2(A) + P_3^0(A) \tag{3.2.4}$$

and get

$$\begin{aligned} 2\partial_t^\perp A_{ij} &= -\nabla_i^\perp \nabla_j^\perp \Delta^\perp \vec{\mathbf{H}} - \nabla_i^\perp \nabla_j^\perp P_3^0(A) \\ &\quad + A * A * \left(P_1^2(A) + P_3^0(A)\right) \\ &= -\nabla_i^\perp \nabla_j^\perp \Delta^\perp \vec{\mathbf{H}} + P_3^2(A) + P_5^0(A). \end{aligned} \tag{3.2.5}$$

To deal with the leading term on the right-hand side in (3.2.5), we recall by [dC, Section 6, Proposition 3.1 and 3.4] the Mainardi-Codazzi equation

$$\nabla_i^\perp A_{jk} = \nabla_j^\perp A_{ki},$$

the Gauß equation

$$R_{ijkl} = \langle A_{ik}, A_{jl}\rangle - \langle A_{il}, A_{jk}\rangle = A * A,$$

and the Ricci equation for a normal vector ϕ

$$\begin{aligned} R_{ij}^\perp \phi &:= \nabla_i^\perp \nabla_j^\perp \phi - \nabla_j^\perp \nabla_i^\perp \phi \\ &= g^{rs} A_{rj} \langle A_{si}, \phi\rangle - g^{rs} A_{ri} \langle A_{sj}, \phi\rangle \\ &= g^{rs} A_{rj}^0 \langle A_{si}^0, \phi\rangle - g^{rs} A_{ri}^0 \langle A_{sj}^0, \phi\rangle \\ &= A * A * \phi. \end{aligned}$$

The Gauß equation yield for a n-covariant, m-contravariant tensor X_β^α

$$R_{ij} X_\beta^\alpha := \nabla_i \nabla_j X_\beta^\alpha - \nabla_j \nabla_i X_\beta^\alpha = g^{\alpha\gamma} R_{ij\gamma k} X_\beta^k + g^{kl} R_{ij\beta k} X_l^\alpha = A * A * X.$$

For a normal, m-covariant tensor ϕ_α, we put

$$R_{ij}^\perp \phi_\alpha := \nabla_i^\perp \nabla_j^\perp \phi_\alpha - \nabla_j^\perp \nabla_i^\perp \phi_\alpha$$

and calculate for a m-contravariant tensor X^α

$$R_{ij}^\perp (\phi_\alpha X^\alpha) = (R_{ij}^\perp \phi_\alpha) X^\alpha + \phi_\alpha R_{ij} X^\alpha.$$

Using the Gauß and the Ricci equation, we get

$$A * A * (\phi_\alpha X^\alpha) = R_{ij}^\perp (\phi_\alpha X^\alpha) = (R_{ij}^\perp \phi_\alpha) X^\alpha + \phi_\alpha\ A * A * X^\alpha,$$

hence

$$R_{ij}^\perp \phi_\alpha = A * A * \phi. \tag{3.2.6}$$

We calculate for a normal, m-covariant tensor $\phi_{i\alpha}$ with α a $m-1$-index that

$$\begin{aligned} g^{kl} \nabla_i^\perp \nabla_k^\perp \phi_{l\alpha} - \Delta^\perp \phi_{i\alpha} &= g^{kl} \nabla_i^\perp \nabla_k^\perp \phi_{l\alpha} - g^{kl} \nabla_k^\perp \nabla_l^\perp \phi_{i\alpha} \\ &= g^{kl} \nabla_i^\perp \nabla_k^\perp \phi_{l\alpha} - g^{kl} \nabla_k^\perp \nabla_i^\perp \phi_{l\alpha} \\ &\quad + g^{kl} \nabla_k^\perp \nabla_i^\perp \phi_{l\alpha} - g^{kl} \nabla_k^\perp \nabla_l^\perp \phi_{i\alpha} \\ &= g^{kl} (R_{ik}^\perp \phi)_{l\alpha} + g^{kl} \nabla_k^\perp \nabla_i^\perp \phi_{l\alpha} - g^{kl} \nabla_k^\perp \nabla_l^\perp \phi_{i\alpha}. \end{aligned}$$

Putting

$$\psi_{il\alpha} := \nabla_l^\perp \phi_{i\alpha} - \nabla_i^\perp \phi_{l\alpha}$$

and using (3.2.6), we obtain

$$g^{kl} \nabla_i^\perp \nabla_k^\perp \phi_{l\alpha} - \Delta^\perp \phi_{i\alpha} = A * A * \phi - g^{kl} \nabla_k^\perp \psi_{li\alpha}, \tag{3.2.7}$$

see [KuSch02, Lemma 2.1 (2.9)].

For $\phi = A$, we get $\psi = 0$ by the Mainardi-Codazzi equation, hence

$$\begin{aligned} \Delta^\perp A_{ij} &= g^{kl} \nabla_i^\perp \nabla_k^\perp A_{lj} + A * A * A \\ &= g^{kl} \nabla_i^\perp \nabla_j^\perp A_{kl} + A * A * A \\ &= \nabla_i^\perp \nabla_j^\perp \vec{\mathbf{H}} + A * A * A, \end{aligned} \tag{3.2.8}$$

which is Simmons' identity.

Applying (3.2.7) to $\nabla^\perp \phi$, we calculate

$$\psi_{ij\alpha} := \nabla_j^\perp \nabla_i^\perp \phi_\alpha - \nabla_i^\perp \nabla_j^\perp \phi_\alpha = (R_{ji}^\perp \phi)_\alpha = A * A * \phi,$$

hence

$$\nabla^\perp \Delta^\perp \phi - \Delta^\perp \nabla^\perp \phi = A * A * \nabla^\perp \phi + A * \nabla^\perp A * \phi. \tag{3.2.9}$$

Applying this to the leading term on the right-hand side in (3.2.5), we continue

$$\begin{aligned} \nabla^{\perp,2} \Delta^\perp \vec{\mathbf{H}} &= \nabla^\perp \Big(\Delta^\perp \nabla^\perp \vec{\mathbf{H}} + A * A * \nabla^\perp \vec{\mathbf{H}} + A * \nabla^\perp A * \vec{\mathbf{H}} \Big) \\ &= \Delta^\perp \nabla^{\perp,2} \vec{\mathbf{H}} + A * A * \nabla^{\perp,2} \vec{\mathbf{H}} + A * \nabla^\perp A * \nabla^\perp \vec{\mathbf{H}} \\ &\quad + \nabla^\perp \Big(A * A * \nabla^\perp \vec{\mathbf{H}} + A * \nabla^\perp A * \vec{\mathbf{H}} \Big) \\ &= \Delta^\perp \nabla^{\perp,2} \vec{\mathbf{H}} + A * A * \nabla^{\perp,2} A + A * \nabla^\perp A * \nabla^\perp A \\ &= \Delta^\perp \nabla^{\perp,2} \vec{\mathbf{H}} + P_3^2(A). \end{aligned}$$

Now using Simmons' identity (3.2.8), we get

$$\Delta^\perp \nabla_{ij}^{\perp,2} \vec{\mathbf{H}} = \Delta^\perp \Big(\Delta^\perp A_{ij} + A * A * A \Big) = \Delta^{\perp,2} A_{ij} + P_3^2(A)$$

and

$$\nabla_{ij}^{\perp,2}\Delta^{\perp}\vec{\mathbf{H}} = \Delta^{\perp,2}A_{ij} + P_3^2(A).$$

Plugging into (3.2.5) yields

$$\partial_t^{\perp}A_{ij} + \frac{1}{2}\Delta^{\perp,2}A_{ij} = P_3^2(A) + P_5^0(A). \tag{3.2.10}$$

Similarly differentiating, we get by induction the evolution equation for the higher derivatives

$$\partial_t^{\perp}\nabla^{\perp,m}A_{ij} + \frac{1}{2}\Delta^{\perp,2}\nabla^{\perp,m}A_{ij} = P_3^{m+2}(A) + P_5^m(A). \tag{3.2.11}$$

Now we differentiate $\int |A|^2\,\mathrm{d}\mu$ and get using (3.2.4), (3.2.10) and (1.3.2), (1.3.3)

$$\begin{aligned}
\frac{1}{2}\frac{d}{dt}\int_{\Sigma} |A|^2\,\mathrm{d}\mu_f = &\int_{\Sigma} g^{ik}g^{jl}\langle\partial_t^{\perp}A_{ij}, A_{kl}\rangle\,\mathrm{d}\mu_f \\
&+ \int_{\Sigma} \partial_t g^{ik}g^{jl}\langle A_{ij}, A_{kl}\rangle\,\mathrm{d}\mu_f \\
&+ \frac{1}{2}\int_{\Sigma} g^{ik}g^{jl}\langle A_{ij}, A_{kl}\rangle\,\mathrm{d}\partial_t\mu_f \\
= &-\frac{1}{2}\int_{\Sigma} g^{ik}g^{jl}\langle\Delta^{\perp,2}A_{ij}, A_{kl}\rangle\,\mathrm{d}\mu_f \\
&- \int_{\Sigma} g^{ik}g^{jl}\langle P_3^2(A) + P_5^0(A), A_{kl}\rangle\,\mathrm{d}\mu_f \\
&+ 2\int_{\Sigma} g^{ir}g^{sk}g^{jl}\langle A_{rs}, V\rangle\langle A_{ij}, A_{kl}\rangle\,\mathrm{d}\mu_f \\
&- \frac{1}{2}\int_{\Sigma} g^{ik}g^{jl}\langle A_{ij}, A_{kl}\rangle\langle\vec{\mathbf{H}}, V\rangle\mu_f \\
= &-\frac{1}{2}\int_{\Sigma} |\Delta^{\perp}A|^2\,\mathrm{d}\mu_f + \int_{\Sigma} (P_4^2(A) + P_6^0(A))\,\mathrm{d}\mu_f.
\end{aligned}$$

Next using (3.2.9)

$$\begin{aligned}\int_\Sigma |\nabla^{\perp,2}A|^2\,d\mu_f &= -\int_\Sigma \langle \Delta^\perp \nabla^\perp A, \nabla^\perp A\rangle\,d\mu_f \\ &= -\int_\Sigma \langle \nabla^\perp \Delta^\perp A, \nabla^\perp A\rangle\,d\mu_f \\ &\quad + \int_\Sigma \langle A * A * \nabla^\perp A, \nabla^\perp A\rangle\,d\mu_f \\ &= \int_\Sigma |\Delta^\perp A|^2\,d\mu_f + \int_\Sigma P_4^2(A)\,d\mu_f,\end{aligned}$$

and we get

$$\frac{d}{dt}\int_\Sigma |A|^2\,d\mu_f + \int_\Sigma |\nabla^{\perp,2}A|^2\,d\mu_f = \int_\Sigma (P_4^2(A) + P_6^0(A))\,d\mu_f.$$

Clearly,

$$\begin{aligned}\Big|\int_\Sigma P_4^2(A)\,d\mu_f\Big| &= \Big|\int_\Sigma (A * A * A * \nabla^{\perp,2}A + A * A * \nabla^\perp A * \nabla^\perp A)\,d\mu_f\Big| \\ &\leq \frac{1}{2}\int_\Sigma |\nabla^{\perp,2}A|^2\,d\mu_f + C\int_\Sigma (|A|^6 + |A|^2\,|\nabla^\perp A|^2)\,d\mu_f,\end{aligned}$$

and we get

$$\begin{aligned}\frac{d}{dt}\int_\Sigma |A|^2\,d\mu_f + \frac{1}{2}\int_\Sigma |\nabla^{\perp,2}A|^2\,d\mu_f \\ \leq C\int_\Sigma (|A|^6 + |A|^2\,|\nabla^\perp A|^2)\,d\mu_f.\end{aligned} \tag{3.2.12}$$

Next we use the Michael-Simon-Sobolev inequality for embedded or immersed manifolds in Euclidean space, see [MiSi73], which reads for the immersion $f : \Sigma \to \mathbb{R}^n$ and any $u \in C^1(\Sigma)$ with compact support

$$\int_\Sigma u^2\,d\mu_f \leq C_n\Big(\int_\Sigma |\nabla u|\,d\mu_f + \int_\Sigma |\vec{\mathbf{H}}|\,|u|\,d\mu_f\Big)^2, \tag{3.2.13}$$

where C_n depends only n and not on Σ. Applying this to $u = |A|\,|\nabla^\perp A|$ and $u = |A|^3$, we get using Kato's inequality $|\nabla|\phi|| \le |\nabla\phi|$ that

$$
\begin{aligned}
&\int_\Sigma |A|^2\,|\nabla^\perp A|^2\,\mathrm{d}\mu_f\\
&\quad\le C_n\Big(\int_\Sigma |\nabla(|A|\,|\nabla^\perp A|)|\,\mathrm{d}\mu_f + \int_\Sigma |\vec{\mathbf{H}}|\,|A|\,|\nabla^\perp A|\,\mathrm{d}\mu_f\Big)^2\\
&\quad\le C\Big(\int_\Sigma (|A|\,|\nabla^{\perp,2}A| + |\nabla^\perp A|^2 + |A|^2\,|\nabla^\perp A|)\,\mathrm{d}\mu_f\Big)^2\\
&\quad\le C\int_\Sigma |A|^2\,\mathrm{d}\mu_f \int_\Sigma |\nabla^{\perp,2}A|^2\,\mathrm{d}\mu_f + C\Big(\int_\Sigma |\nabla^\perp A|^2\,\mathrm{d}\mu_f\Big)^2\\
&\qquad + C\Big(\int_\Sigma |A|^4\,\mathrm{d}\mu_f\Big)^2
\end{aligned}
$$

and

$$
\begin{aligned}
\int_\Sigma |A|^6\,\mathrm{d}\mu_f &\le C_n\Big(\int_\Sigma |\nabla(|A|^3)|\,\mathrm{d}\mu_f + \int_\Sigma |\vec{\mathbf{H}}|\,|A|^3\,\mathrm{d}\mu_f\Big)^2\\
&\le C\Big(\int_\Sigma (|A|^2\,|\nabla^\perp A| + |A|^4)\,\mathrm{d}\mu_f\Big)^2\\
&\le C\Big(\int_\Sigma |\nabla^\perp A|^2\,\mathrm{d}\mu_f\Big)^2 + C\Big(\int_\Sigma |A|^4\,\mathrm{d}\mu_f\Big)^2.
\end{aligned}
$$

Now

$$
\Big(\int_\Sigma |A|^4\,\mathrm{d}\mu_f\Big)^2 \le \int_\Sigma |A|^2\,\mathrm{d}\mu_f \int_\Sigma |A|^6\,\mathrm{d}\mu_f
$$

and

$$
\begin{aligned}
\int_\Sigma |\nabla^\perp A|^2\,\mathrm{d}\mu_f &= -\int_\Sigma A\,\Delta^\perp A\,\mathrm{d}\mu_f\\
&\le \Big(\int_\Sigma |A|^2\,\mathrm{d}\mu_f \int_\Sigma |\Delta^\perp A|^2\,\mathrm{d}\mu_f\Big)^{1/2},
\end{aligned}
$$

hence

$$
\int_\Sigma (|A|^6 + |A|^2\,|\nabla^\perp A|^2)\,\mathrm{d}\mu_f \le C\int_\Sigma |A|^2\,\mathrm{d}\mu_f \int_\Sigma (|\nabla^{\perp,2}A|^2 + |A|^6)\,\mathrm{d}\mu_f. \tag{3.2.14}
$$

Plugging this into (3.2.12), we cannot absorb the term $\int_\Sigma |\nabla^\perp A|^2 \, d\mu_f$, as the factor in front satisfies by (1.1.9) and (2.1.18)

$$\int_\Sigma |A|^2 \, d\mu_f = 4\mathcal{W}(f) + 8\pi(p(\Sigma) - 1) \geq 8\pi$$

and can therefore not be arbitrarily small. Therefore we have to localize (3.2.12) and (3.2.14). We choose $\tilde{\gamma} \in C^2(\mathbb{R}^n)$ with $\| \tilde{\gamma} \|_{C^2(\mathbb{R}^n)} \leq C$, put $\gamma = \tilde{\gamma} \circ f$ and get

$$\begin{aligned}
&\frac{d}{dt} \int_\Sigma |A|^2 \gamma^4 \, d\mu_f + \frac{1}{2} \int_\Sigma |\nabla^{\perp,2} A|^2 \gamma^4 \, d\mu_f \\
&\quad \leq C \int_\Sigma (|A|^6 + |A|^2 \, |\nabla^\perp A|^2) \gamma^4 \, d\mu_f + C \int_{[\gamma>0]} |A|^2 \, d\mu_f
\end{aligned}$$

and

$$\begin{aligned}
&\int_\Sigma (|A|^6 + |A|^2 \, |\nabla^\perp A|^2) \gamma^4 \, d\mu_f \\
&\quad \leq C \int_{[\gamma>0]} |A|^2 \, d\mu_f \int_\Sigma (|\nabla^{\perp,2} A|^2 + |A|^6) \gamma^4 \, d\mu_f + C\Big(\int_{[\gamma>0]} |A|^2 \, d\mu_f \Big)^2.
\end{aligned}$$

Assuming

$$\int_{[\gamma>0]} |A|^2 \, d\mu_f \leq \tilde{\varepsilon}_0$$

for $\tilde{\varepsilon}_0$ small enough, depending on C, we obtain

$$\begin{aligned}
&\frac{d}{dt} \int_\Sigma |A|^2 \gamma^4 \, d\mu_f + \frac{1}{4} \int_\Sigma |\nabla^{\perp,2} A|^2 \gamma^4 \, d\mu_f \\
&\quad \leq C\Big(\int_{[\gamma>0]} |A|^2 \, d\mu_f \Big)^2.
\end{aligned} \tag{3.2.15}$$

To prove the theorem, we may assume after parabolic rescaling that $\varrho = 1$. We choose $\chi_{B_{1/2}(x)} \leq \tilde{\gamma} \leq \chi_{B_1(x)}$ with $\| \tilde{\gamma} \|_{C^2(\mathbb{R}^n)} \leq C_n$. By a simple covering argument, we get

$$\sup_{x \in \mathbb{R}^n} \int_{B_1(x)} |A|^2 \, d\mu_f \leq \Gamma_n \sup_{x \in \mathbb{R}^n} \int_{B_{1/2}(x)} |A|^2 \, d\mu_f$$

for some $1 \leq \Gamma_n < \infty$. Putting

$$\varepsilon(t) := \sup_{x \in \mathbb{R}^n} \int\limits_{B_1(x)} |A_t|^2 \, d\mu_{f_t},$$

we define

$$t_0 := \sup\{0 \leq t \leq T \mid \varepsilon(\tau) \leq 3\Gamma_n \varepsilon \text{ for all } 0 \leq \tau \leq t\} \qquad (3.2.16)$$

and see $0 < t_0 \leq T$ by continuity of $\varepsilon(.)$ and by (3.2.1). For $3\Gamma_n\varepsilon_0 \leq \tilde{\varepsilon}_0(C_n)$, we get by (3.2.15) that

$$\int\limits_{B_{1/2}(x)} |A_t|^2 \, d\mu_{f_t} \leq \int\limits_{B_1(x)} |A_0|^2 \, d\mu_{f_0} + C_n(\Gamma_n\varepsilon)^2 t \quad \text{for all } x \in \mathbb{R}^n, 0 \leq t < t_0,$$

hence

$$\varepsilon(t) \leq \Gamma_n(\varepsilon + C_n(\Gamma_n\varepsilon)^2 t_0) \quad \text{for all } 0 \leq t < t_0.$$

Assuming $\Gamma_n\varepsilon_0 \leq 1$, we get for $t_0 \leq c_0(n)$ that

$$\varepsilon(t) \leq 2\Gamma_n\varepsilon \quad \text{for all } 0 \leq t < t_0.$$

If we had $t_0 < \min(T, c_0)$, we conclude by continuity that $\varepsilon(t_0) = 3\Gamma_n\varepsilon$ contradicting the above estimate. Therefore $t_0 \geq \min(T, c_0)$, and we get

$$\int\limits_{B_1(x)} |A|^2 \, d\mu_f \leq 2\Gamma_n\varepsilon \quad \text{for all } x \in \mathbb{R}^n, 0 \leq t < \min(T, c_0)$$

which is (3.2.3).

In case $T \leq c_0$, we have $T = \min(T, c_0) \leq t_0 \leq T$, hence $t_0 = T \leq c_0$. Then $\varepsilon(t) \leq 2\Gamma_n\varepsilon$ for all $0 \leq t < T$. By applying the evolution equations for the higher derivatives (3.2.11), we can estimate all derviatives of A and f for $0 \leq t < T$, and f can be extended smoothly to $\Sigma \times [0, T]$. Moreover by (1.3.2)

$$|\partial_t g_{ij}| \leq 2|A|\,|V| \leq C,$$

hence

$$\int\limits_0^T \| \, |\partial_t g_{ij}| \, \|_{L^\infty(\Sigma)} \, dt < \infty,$$

and by [Ha82] Lemma 14.2 the metrics $g(t)$ are uniformly equivalent. Therefore $f(T)$ is a smooth immersion and by local existence f can be extended to a Willmore flow for $0 \leq t < T + \delta$ for some $\delta > 0$. This contradicts the maximality of T, hence $T \geq c_0$ which is (3.2.2) and the theorem is proved. □

Actually singularities may occur for the Willmore flow that is there is no global in time existence or in case of global existence there is no convergence for $t \to \infty$. The following data when rotated at the x-axis

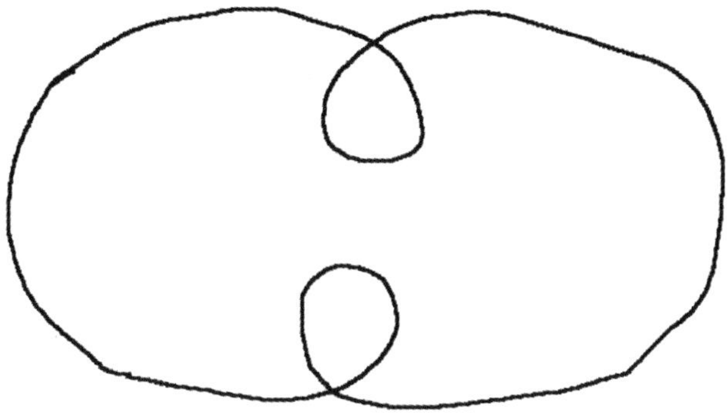

were proposed in [MaSi02] to develop a singularity and there numerical evidence was given. This was rigorously proved in [Bl09]. Still it is open whether the singularity develops in finite time or at infinity.

3.3 Blow up

To study the singulartities in greater detail, we set up a blow up procedure. With the Michael-Simon-Sobolev inequality and the absorption technique in the above theorem, one can prove interior estimate stated in the next theorem.

Theorem 3.3.1 ([KuSch01] Theorem 3.5). *Let* $f : \Sigma \times [0, T[\to \mathbb{R}^n$ *be a normal Willmore flow* $\partial_t f_t + \delta\mathcal{W}(f_t) = 0$ *of a closed surface* Σ *with* $0 < T \leq \infty$ *satisfying*

$$\sup_{0 \leq t < T} \int_{B_\varrho(0)} |A_t|^2 \, d\mu_{f_t} \leq \varepsilon \leq \varepsilon_1(n) \tag{3.3.1}$$

for some $\varrho > 0$ *and appropriate* $\varepsilon_1(n) > 0$.

Then we have for $0 < t < c_0(n)\varrho^4, T$ *and for any* $k \in \mathbb{N}_0$

$$\begin{aligned} \| \nabla^{\perp,k} A_t \|_{L^\infty(B_{1/2}(0))} &\leq C_k(n)\sqrt{\varepsilon}\, t^{-(k+1)/4}, \\ \| \nabla^{\perp,k} A_t \|_{L^2(B_{1/2}(0))} &\leq C_k(n)\sqrt{\varepsilon}\, t^{-k/4}. \end{aligned} \tag{3.3.2}$$

We define the critical radius

$$r_t := \sup\{\varrho > 0 \mid \forall x \in \mathbb{R}^n : \int_{B_\varrho(x)} |A_{f_t}|^2 \, d\mu_{f_t} \leq \varepsilon_0 \} \tag{3.3.3}$$

for $0 \le t < T$ and see

$$\int_{B_{r_t}(x)} |A_{f_t}|^2 \, d\mu_{f_t} \le \varepsilon_0 \quad \forall x \in \mathbb{R}^n. \tag{3.3.4}$$

Moreover there exists $x_t \in \mathbb{R}^n$ not necessarily unique with

$$\int_{B_{r_t}(x_t)} |A_{f_t}|^2 \, d\mu_{f_t} \ge \varepsilon_0. \tag{3.3.5}$$

If T is the maximal existence time, we get by Theorem 3.2.1

$$c_0 r_t^4 < (T - t) \quad \text{for } 0 \le t < T, \tag{3.3.6}$$

in particular

$$r_t \to 0 \quad \text{for } t \nearrow T, \text{ if } T < \infty, \tag{3.3.7}$$

and

$$\int_{B_{r_t}(x)} |A_{f_{t+s}}|^2 \, d\mu_{f_{t+s}} \le C\varepsilon_0 \quad \text{for } 0 \le s \le c_0 r_t^4, 0 \le t < T, x \in \mathbb{R}^n.$$

We choose $0 < \varrho_t \le r_t$, put $f_s^t := \varrho_t^{-1}(f_{t+s\varrho_t^4} - y_t)$ for $0 \le s \le c_0$, $y_t \in \mathbb{R}^n$, and see

$$\int_{B_1(x)} |A_{f_s^t}|^2 \, d\mu_{f_s^t} \le \int_{B_{r_t/\varrho_t}(x)} |A_{f_s^t}|^2 \, d\mu_{f_s^t} \le C\varepsilon_0 \quad \text{for } 0 \le s \le c_0, 0 \le t < T, x \in \mathbb{R}^n, \tag{3.3.8}$$

hence for $C\varepsilon_0 \le \varepsilon_1$ by Theorem 3.3.1 that

$$\|\nabla^k A_{f_s^t}\|_{L^\infty(\Sigma)} \le C_k(n) s^{-(k+1)/4} \text{ for } 0 \le s \le c_0, 0 \le t < T, k \in \mathbb{N}_0. \tag{3.3.9}$$

Moreover by (2.1.15) and $\mathcal{W}(f_t) \le \mathcal{W}(f_0)$

$$\varrho^{-2} \mu_{f_s^t}(B_\varrho(x)) \le C\mathcal{W}(f_0) \quad \forall x \in \mathbb{R}^n. \tag{3.3.10}$$

Then as in [KuSch01, Section 4] for appropriate subsequences $t_j \to T$, the flows $(f_s^{t_j})_{s \in]0,c_0]}$ converge after appropriate reparametrization smoothly on compact subsets of $\mathbb{R}^n$ to a Willmore flow $(f_s^\infty)_{s \in]0,c_0]} : \Sigma^\infty \to \mathbb{R}^n$ of an open two-manifold Σ^∞ without boundary, which is possibly not connected. We calculate as in [KuSch01, Section 4] using (1.3.6)

$$\int_0^{c_0}\int_{\Sigma^\infty} |\partial_t f_s^\infty|^2 \, d\mu_{f_s^\infty} \, ds \le \liminf_{j\to\infty} \int_0^{c_0}\int_{\Sigma} |\partial_t f_s^{t^j}|^2 \, d\mu_{f_s^{t^j}} \, ds$$
$$= \liminf_{j\to\infty} 2\Big(\mathcal{W}(f_{t_j}) - \mathcal{W}(f_{t_j + c_0\varrho_{t_j}^4})\Big) = 0,$$

as $\mathcal{W}(f_t)$ is non-increasing and therfore converges for $t \to T$. We conclude that the limit flow (f_s^∞) is stationary, and $f^\infty := f_s^\infty$ is a smooth proper Willmore immersion.

So far f^∞ can be trivial being planes or even $\Sigma^\infty = \emptyset$. To apply this blow up procedure, one has to ensure that f^∞ is non-trivial.

We call f^∞ a limit under the blow up procedure, and more precisely, we call f^∞ a blow up for $\varrho_{t_j} \to 0$, a blow down for $\varrho_{t_j} \to \infty$ and a limit under translations for $\varrho_{t_j} \to 1$. Clearly by (3.3.7), the case $\varrho_{t_j} \to r \in \{1, \infty\}$ only occur, if $T = \infty$.

If Σ^∞ contains a compact component, then $\Sigma^\infty \cong \Sigma$, see [KuSch01] Lemma 4.3. Actually blow ups and blow downs are never compact, see [ChFaSch09].

3.4 Convergence to a sphere

According to a theorem of Codazzi the only closed surfaces which are totally umbilic, that is $A^0 \equiv 0$, are the round spheres. In this last section, we prove if the inital data are close to a round sphere then the Willmore flow converges to a round sphere.

Theorem 3.4.1 ([KuSch01, Theorem 5.1]). *There exists $\varepsilon_2(n) > 0$ such that if*

$$\int_\Sigma |A_0^0|^2 \, d\mu_{f_0} \le \varepsilon_2(n)$$

then the Willmore flow (f_t) with initial data f_0 exists globally in time and converges for $t \to \infty$ exponentially to a round sphere.

Firstly, for a immersion $f : \Sigma \to \mathbb{R}^n$ of a closed surface Σ we see by (1.1.9) and (2.1.18)

$$\int_\Sigma |A_f^0|^2 \, d\mu_f = 2\mathcal{W}(f) - 4\pi\chi(\Sigma) \ge 4\pi(2 - \chi(\Sigma)),$$

hence for $\varepsilon_2(n) < 4\pi$ in the above theorem, we get $\chi(\Sigma) > 1$, hence $\Sigma \cong S^2$, see for example [Ms77, Theorems 5.1 and 8.2].

Instead of getting (3.2.15) by localization, one can use the smallness of $\int_\Sigma |A^0_f|^2 \, d\mu_f$ to obtain a tracefree curvature estimate.

Proposition 3.4.1 ([KuSch01, Proposition 2.6]). *Let $f : \Sigma \to \mathbb{R}^n$ be a proper Willmore immersion of an open surface Σ satisfying*

$$\int_\Sigma |A^0_f|^2 \, d\mu_f \leq \varepsilon_1,$$
$$\lim_{\varrho \to \infty} \varrho^{-4} \int_{B_\varrho(0)} |A_f|^2 \, d\mu_f = 0,$$

in particular if $\int_\Sigma |A_f|^2 \, d\mu_f < \infty$, then

$$\int_\Sigma (|\nabla^{\perp,2} A|^2 + |A|^2 \, |\nabla^\perp A|^2 + |A|^4 |A^0|^2) \, d\mu_f \leq C_n \int_\Sigma |\delta \mathcal{W}(f)|^2 \, d\mu_f.$$

This proposition implies the following gap lemma.

Lemma 3.4.1 (Gap Lemma [KuSch01, Theorem 2.7]). *Under the assumptions of the previous proposition, if f is Willmore, then f parametrises on each connected component of Σ an embedded plane or a round sphere.*

Proof. f is Willmore if and only if $\delta \mathcal{W}(f) = 0$, hence $A^0 \equiv 0$, and f maps each connected component of Σ into an embedded plane or round sphere by Codazzi's theorem, say $f : \Sigma_0 \to M$ for $\Sigma_0 \subseteq \Sigma$ a connected componenet and $M = P$ a plane or $= S^2$. Then f being an immersion is a local diffeomorphism, hence $f(\Sigma_0) \subseteq M$ is open. On the other hand, $f(\Sigma_0)$ is closed, as f is proper, hence $f(\Sigma_0) = M$, as M is connected.

Since f is a local diffeomorphism, the fiber $f^{-1}(x)$ for $x \in M$ is a discrete set, and as f is proper, $f^{-1}(x)$ is compact, hence finite. Then f being a local diffeomorphism is a covering map, and hence is a diffeomorphism, as M is simply connected. □

We use this to prove the following estimate for Theorem 3.4.1.

Proposition 3.4.2. *Under the assumptions of Theorem* 3.4.1

$$\begin{aligned} &\int_0^T \int_\Sigma (|\nabla^{\perp,2} A_t|^2 + |A_t|^2 \, |\nabla^\perp A_t|^2 + |A_t|^4 |A^0_t|^2) \, d\mu_{f_t} \, dt \\ &\leq C_n \int_\Sigma |A^0_{f_0}|^2 \, d\mu_{f_0} \end{aligned} \tag{3.4.1}$$

and

$$\mu_\infty := \lim_{t\to\infty} \mu_{f_t}(\mathbb{R}^n) \in]0, \infty[\tag{3.4.2}$$

exists, and this limit is finite and positive.

Proof. By (1.1.9) and (1.3.6), we see for $0 \le s \le t < T$

$$\int_s^t \int_\Sigma |\delta\mathcal{W}(f_\tau)|^2 \, \mathrm{d}\mu_{f_\tau} \, \mathrm{d}\tau = \int_\Sigma |A^0_{f_s}|^2 \, \mathrm{d}\mu_{f_s} - \int_\Sigma |A^0_{f_t}|^2 \, \mathrm{d}\mu_{f_t}$$

and by Proposition 3.4.1

$$\begin{aligned} &\int_s^t \int_\Sigma (|\nabla^{\perp,2} A_t|^2 + |A_t|^2 \, |\nabla^\perp A_t|^2 + |A_t|^4 |A^0_t|^2) \, \mathrm{d}\mu_{f_t} \, \mathrm{d}t \\ &\le C_n \int_s^t \int_\Sigma |\delta\mathcal{W}(f_t)|^2 \, \mathrm{d}\mu_{f_t} \, \mathrm{d}t \\ &= C_n \Big(\int_\Sigma |A^0_{f_s}|^2 \, \mathrm{d}\mu_{f_s} - \int_\Sigma |A^0_{f_t}|^2 \, \mathrm{d}\mu_{f_t} \Big) \\ &\le C_n \int_\Sigma |A^0_{f_0}|^2 \, \mathrm{d}\mu_{f_0} \le C_n \varepsilon_2, \end{aligned} \tag{3.4.3}$$

which yields (3.4.1).

Next by (1.3.3) and (1.3.8)

$$\begin{aligned} \frac{d}{dt} \mu_{f_t}(\mathbb{R}^n) &= - \int_\Sigma \langle \vec{\mathbf{H}}, \partial f_t \rangle \, \mathrm{d}\mu_{f_t} \\ &= - \int_\Sigma |\nabla^\perp \vec{\mathbf{H}}|^2 \, \mathrm{d}\mu_{f_t} + \int_\Sigma \langle Q(A^0)\vec{\mathbf{H}}, \vec{\mathbf{H}} \rangle \, \mathrm{d}\mu_{f_t}. \end{aligned} \tag{3.4.4}$$

By the Michael-Simon-Sobolev inequality (3.2.13) applied to $u=|\vec{\mathbf{H}}|\,|A^0|$ and Kato's inequality, we get

$$\int_\Sigma \langle Q(A^0)\vec{\mathbf{H}}, \vec{\mathbf{H}}\rangle \, d\mu_f \le \int_\Sigma |\vec{\mathbf{H}}|^2\, |A^0|^2 \, d\mu_f$$

$$\le C_n\Big(\int_\Sigma |\nabla^\perp \vec{\mathbf{H}}|\, |A^0| \, d\mu_f + \int_\Sigma |\vec{\mathbf{H}}|\, |\nabla^\perp A^0| \, d\mu_f + \int_\Sigma |\vec{\mathbf{H}}|^2\, |A^0| \, d\mu_f\Big)^2$$

$$\le C_n \int_\Sigma |A^0|^2 \, d\mu_f \int_\Sigma |\nabla^\perp \vec{\mathbf{H}}|^2 \, d\mu_f$$

$$+ C_n \mu_f(\mathbb{R}^n)\Big(\int_\Sigma |A|^2\, |\nabla^\perp A|^2 \, d\mu_f + \int_\Sigma |A|^4\, |A^0|^2 \, d\mu_f\Big),$$

and for $C_n \varepsilon_2(n) \le 1$

$$-\int_\Sigma |\nabla^\perp \vec{\mathbf{H}}|^2 \, d\mu_f + \int_\Sigma \langle Q(A^0)\vec{\mathbf{H}}, \vec{\mathbf{H}}\rangle \, d\mu_f$$

$$\le C_n \mu_f(\mathbb{R}^n)\Big(\int_\Sigma |A|^2\, |\nabla^\perp A|^2 + |A|^4\, |A^0|^2 \, d\mu_f\Big).$$

Then by (3.4.3), (3.4.4) and Gronwall's lemma

$$\mu_{f_t}(\mathbb{R}^n) \le \mu_{f_0}(\mathbb{R}^n) \exp\Big(C_n \int_0^T \int_\Sigma (|A_t|^2\, |\nabla^\perp A_t|^2 + |A_t|^4 |A_t^0|^2) \, d\mu_{f_t} \, dt\Big)$$

$$\le e^{C_n \varepsilon_2} \mu_{f_0}(\mathbb{R}^n).$$

On the other hand by (1.3.7)

$$\langle Q(A^0)\vec{\mathbf{H}}, \vec{\mathbf{H}}\rangle = g^{ik} g^{jl} \langle A^0_{ij}, \vec{\mathbf{H}}\rangle \langle A^0_{kl}, \vec{\mathbf{H}}\rangle \ge 0,$$

and again by Michael-Simon-Sobolev inequality (3.2.13) and Kato's inequality

$$\int_\Sigma |\nabla^\perp \vec{\mathbf{H}}|^2 \, d\mu_f \le C_n\Big(\int_\Sigma |\nabla^{\perp,2} \vec{\mathbf{H}}| \, d\mu_f + \int_\Sigma |\vec{\mathbf{H}}|\, |\nabla^\perp \vec{\mathbf{H}}| \, d\mu_f\Big)^2$$

$$\le C_n \mu_f(\mathbb{R}^n) \int_\Sigma (|\nabla^{\perp,2} \vec{\mathbf{H}}|^2 + |\vec{\mathbf{H}}|^2\, |\nabla^\perp \vec{\mathbf{H}}|^2) \, d\mu_f,$$

hence by (3.4.4)

$$\frac{d}{dt}\mu_{f_t}(\mathbb{R}^n) \geq -C_n \mu_{f_t}(\mathbb{R}^n) \int_\Sigma (|\nabla^{\perp,2}\vec{\mathbf{H}}|^2 + |\vec{\mathbf{H}}|^2\, |\nabla^\perp \vec{\mathbf{H}}|^2)\, \mathrm{d}\mu_f.$$

Again by (3.4.3)

$$\begin{aligned}
&\int_s^t \int_\Sigma (|\nabla^{\perp,2}\vec{\mathbf{H}}|^2 + |\vec{\mathbf{H}}|^2\, |\nabla^\perp \vec{\mathbf{H}}|^2)\, \mathrm{d}\mu_f\, \mathrm{d}\tau \\
&\leq \int_s^t \int_\Sigma (|\nabla^{\perp,2} A|^2 + |A|^2\, |\nabla^\perp A|^2)\, \mathrm{d}\mu_f\, \mathrm{d}\tau \\
&\leq C_n\Big(\int_\Sigma |A^0_{f_s}|^2\, \mathrm{d}\mu_{f_s} - \int_\Sigma |A^0_{f_t}|^2\, \mathrm{d}\mu_{f_t} \Big),
\end{aligned}$$

we get by Gronwall's lemma

$$\exp\Big(-C_n \int_\Sigma |A^0_{f_t}|^2\, \mathrm{d}\mu_{f_t}\Big)\mu_{f_t}(\mathbb{R}^n) \geq \exp\Big(-C_n \int_\Sigma |A^0_{f_s}|^2\, \mathrm{d}\mu_{f_s}\Big)\mu_{f_s}(\mathbb{R}^n)$$

and $t \mapsto \exp(-C_n \int_\Sigma |A^0_{f_t}|^2\, \mathrm{d}\mu_{f_t})\mu_{f_t}(\mathbb{R}^n)$ is non-decreasing. In particular its limit for $t \to \infty$ exists, and as $t \mapsto \int_\Sigma |A^0_{f_t}|^2\, \mathrm{d}\mu_{f_t}$ is non-increasing, the limit

$$\mu_\infty := \lim_{t\to\infty} \mu_{f_t}(\mathbb{R}^n) \in [0, \infty]$$

exists. We see

$$\mu_\infty \geq e^{-C_n \varepsilon_2} \mu_{f_0}(\mathbb{R}^n) > 0$$

and by above

$$\mu_\infty \leq e^{C_n \varepsilon_2} \mu_{f_0}(\mathbb{R}^n) < \infty,$$

concluding the proof. □

Proposition 3.4.3. *Under the assumptions of Theorem* 3.4.1 *there exists* $r > 0$ *such that*

$$\int_{B_r(x)} |A_t|^2\, \mathrm{d}\mu_{f_t} \leq \varepsilon_0 \quad \forall 0 \leq t < T, x \in \mathbb{R}^n, \tag{3.4.5}$$

in particular the flows exists global in time.

Proof. Let $f_t : \Sigma \to \mathbb{R}^n$ be a Willmore flow $\partial_t f_t + \delta \mathcal{W}(f_t) = 0$ on the maximal existence interval $[0, T[$ with $0 < T \leq \infty$ satisfying

$$\int_\Sigma |A_0^0|^2 \, d\mu_{f_0} \leq \varepsilon_2(n).$$

If there is no such $r > 0$ then for $0 < \varrho_j \to 0$ there exists $0 \leq \tilde{t}_j < T, x_j \in \mathbb{R}^n$ with

$$\int_{B_{\varrho_j}(x)} |A_{\tilde{t}_j}|^2 \, d\mu_{f_{\tilde{t}_j}} > \varepsilon_0,$$

hence for the critical radius in (3.3.3) that $r_{\tilde{t}_j} \leq \varrho_j$ and

$$\hat{t}_j := \inf\{0 \leq t < T \mid r_t \leq \varrho_j\} \leq \tilde{t}_j < T.$$

Clearly,

$$r_t > \varrho_j \quad \text{for } 0 \leq t < \hat{t}_j,$$

and there is $s_k \searrow \hat{t}_j$ with $r_{s_k} \leq \varrho_j$. Then by (3.3.5) there exists $x_k \in \mathbb{R}^n$ with

$$\int_{\overline{B_{\varrho_j}(x_k)}} |A_{f_{s_k}}|^2 \, d\mu_{f_{s_k}} \geq \varepsilon_0.$$

As $f(\Sigma \times [0, T_0])$ is compact for $0 < T_0 < T$ and $\overline{B_{\varrho_j}(x_k)} \cap f(\Sigma \times [0, T_0]) \neq \emptyset$ for $\hat{t}_j < T_0$ and k large, we get for a subsequence $x_k \to y_j$. Then by smoothness of f on $[0, T[$

$$\int_{\overline{B_{\varrho_j}(y_j)}} |A_{f_{\hat{t}_j}}|^2 \, d\mu_{f_{\hat{t}_j}} \geq \limsup_{k \to \infty} \int_{\overline{B_{\varrho_j}(x_k)}} |A_{f_{s_k}}|^2 \, d\mu_{f_{s_k}} \geq \varepsilon_0.$$

We put $t_j := \hat{t}_j - c_0 \varrho_j^4$. As f is smooth, $\hat{t}_j, t_j \to T$, in particular $t_j > 0$ for j large and $\varrho_j \leq r_{t_j}$. Then performing a blow up by the previous section,

$$f_s^j := \varrho_j^{-1}(f_{t_j + s\varrho_j^4} - y_j) \quad \text{for } s \in]0, c_0]$$

converges for a subsequence after reparametrization to a smooth proper Willmore immersion $f^\infty : \Sigma^\infty \to \mathbb{R}^n$. Clearly,

$$\int_{\Sigma^\infty} |A_{f^\infty}^0|^2 \, d\mu_{f^\infty} \leq \liminf_{j \to \infty} \int_\Sigma |A_{f_{c_0}^j}^0|^2 \, d\mu_{f_{c_0}^j} \leq \int_\Sigma |A_{f_0}^0|^2 \, d\mu_{f_0} \leq \varepsilon_2,$$

$$\int_{\Sigma^\infty} |A_{f^\infty}|^2 \, d\mu_{f^\infty} \leq \liminf_{j \to \infty} \int_\Sigma |A_{f_{c_0}^j}|^2 \, d\mu_{f_{c_0}^j} \leq \int_\Sigma |A_{f_0}|^2 \, d\mu_{f_0} < \infty,$$

hence for $\varepsilon_2 \leq \varepsilon_1$, we get from the Gap Lemma 3.4.1 that f parametrises planes and round spheres.

Moreover

$$\int_{B_1(0)} |A_{f^\infty}|^2 \, d\mu_{f^\infty} \geq \limsup_{j\to\infty} \int_{B_1(0)} |A_{f^j_{c0}}|^2 \, d\mu_{f^j_{c0}}$$

$$= \limsup_{j\to\infty} \int_{B_{\varrho_j}(y_j)} |A_{f_{\hat{t}_j}}|^2 \, d\mu_{f_{\hat{t}_j}} \geq \varepsilon_0,$$

and f^∞ is non-trivial in that sense that $\Sigma^\infty \neq \emptyset$ and f^∞ does not parametizes only planes. Therefore there is at least one component of Σ^∞ over which f^∞ parametrises a round sphere, hence this component is compact. By the remark at the end of Section 3.3, this is the only component and $f^\infty : \Sigma^\infty \overset{\approx}{\longrightarrow} S$ parametrises a round sphere S. In particular $\mu_{f^j_{c0}}(\mathbb{R}^n) \to \mathcal{H}^2(S) < \infty$ and

$$\mu_{\hat{f}_{t_j}}(\mathbb{R}^n) = \varrho_j^2 \mu_{f^j_{c0}}(\mathbb{R}^n) \to 0.$$

But by (3.4.2)

$$\mu_{\hat{f}_{t_j}}(\mathbb{R}^n) \to \mu_\infty > 0$$

which is a contradiction, and hence exists $r > 0$ satisfying (3.4.5). We conclude $r \leq r_t$ for all $0 \leq t < T$, hence $T = \infty$ by (3.3.7), which means global existence. □

Proof of Theorem 3.4.1. By Proposition 3.4.3 and the interior estimates in Theorem 3.3.1, we see

$$\| \nabla^k A_t \|_{L^\infty(\Sigma)} \leq C_k(n) \quad \text{for } 1 \leq t < \infty, k \in \mathbb{N}_0.$$

In the blow up procedure of the previous section, we can choose $\varrho_t := r \leq r_t$ and see that for any $t_j \to \infty, y_j \in \mathbb{R}^n$ there is a subsequence such that $f_{t_j} - y_j$ converges after reparametrization to a smooth proper Willmore immersion $f^\infty : \Sigma^\infty \to \mathbb{R}^n$. As above

$$\int_{\Sigma^\infty} |A^0_{f^\infty}|^2 \, d\mu_{f^\infty} \leq \varepsilon_2, \qquad \int_{\Sigma^\infty} |A_{f^\infty}|^2 \, d\mu_{f^\infty} < \infty,$$

hence for $\varepsilon_2 \leq \varepsilon_1$, we get from the Gap Lemma 3.4.1 that f parametrises planes and round spheres. By (3.4.2)

$$r^2 \mu_{f^\infty}(\mathbb{R}^n) = \lim_{t\to\infty} \mu_{f_t}(\mathbb{R}^n) = \mu_\infty < \infty,$$

and hence there are no planes.

Choosing $y_j \in f_{t_j}(\Sigma)$, we know $0 \in f^\infty(\Sigma^\infty)$, in particular $\Sigma^\infty \neq \emptyset$. Therefore there is at least one component of Σ^∞ over which f^∞ parametrises a round sphere, hence this component is compact. By the remark at the end of Section 3.3, this is the only component and $f^\infty : \Sigma^\infty \xrightarrow{\approx} S$ parametrises a round sphere S. Again by (3.4.2)

$$r^2\mathcal{H}^2(S) = r^2\mu_{f^\infty}(\mathbb{R}^n) \leftarrow \mu_{f_{t_j}}(\mathbb{R}^n) \to \mu_\infty,$$

and although S is not unique, the radius of S is unique be given by $\varrho = \sqrt{\mu_\infty/4\pi}\, r^{-1}$. We conclude

$$\left.\begin{array}{r} |A^0_{f_t}| \to 0, \\ |\vec{\mathbf{H}}_{f_t}| \to \sqrt{4\pi/\mu_\infty}, \end{array}\right\} \quad \text{uniformly for } t \to \infty,$$

in particular

$$|\vec{\mathbf{H}}_{f_t}| \geq c > 0 \quad \text{for } t \geq t_0.$$

Then by Proposition 3.4.1 and (1.1.9) and (1.3.6)

$$\begin{aligned} \partial_t \int_\Sigma |A^0_{f_t}|^2 \, d\mu_{f_t} &= -\int_\Sigma |\delta\mathcal{W}(f_t)|^2 \, d\mu_{f_t} \\ &\leq -C_n \int_\Sigma (|\nabla^{\perp,2} A|^2 + |A|^2\, |\nabla^\perp A|^2 + |A|^4 |A^0|^2) \, d\mu_f \\ &\leq -C_n c^4 \int_\Sigma (|\nabla^{\perp,2} A|^2 + |\nabla^\perp A|^2 + |A^0|^2) \, d\mu_f \end{aligned}$$

for $c \leq 1$, hence by Gronwall's lemma observing $\int_\Sigma |A^0_{f_t}|^2 \, d\mu_{f_t} \to 0$ for $t \to \infty$

$$\int_\Sigma |A^0_{f_t}|^2 \, d\mu_{f_t} + \int_t^\infty (|\nabla^{\perp,2} A_\tau|^2 + |\nabla^\perp A_\tau|^2) \, d\mu_{f_\tau} \, d\tau \leq Ce^{-2\lambda t}$$

for some $C = C_n < \infty$ and $\lambda = \lambda_n > 0$. Doing estimates on the higher derivatives, we obtain

$$\begin{aligned} \| A^0_t \|_{L^\infty(\Sigma)} &\leq Ce^{-\lambda t}, \\ \| \nabla^{\perp,k} A_t \|_{L^\infty(\Sigma)} &\leq C_k e^{-\lambda t} \quad \text{for } k \geq 1, \end{aligned}$$

which yields exponential convergence of f. As we have already seen above, the limit is a round sphere. □

Appendix

3.A Blow up radius

In this appendix, we just remark on the relation between r_t and ϱ_t in the blow up procedure of Section 3.3.

Proposition 3.A.1. *If*

$$\limsup_{j\to\infty} r_{t_j}/\varrho_{t_j} = \infty,$$

then f^∞ *is trivial.*

On the other hand, there exists $t_j \to T$ *such that for* $\varrho_{t_j} = r_{t_j}$ *and appropriate* y_{t_j} *the limit* f^∞ *is non-trivial, that is* f^∞ *does not parametrise only a union of planes, in particular* $\Sigma^\infty \neq \emptyset$.

Proof. We put $R_t = r_t/\varrho_t \geq 1$ and assume for a subsequence $R_{t_j} \to \infty$. We see by (3.3.8) and by lower semicontinuity that

$$\int_{\Sigma^\infty} |A_{f^\infty}|^2 \,\mathrm{d}\mu_{f^\infty} \leq \liminf_{j\to\infty} \int_{B_{R_{t_j}}(x)} |A_{f_s^{t_j}}|^2 \,\mathrm{d}\mu_{f_s^{t_j}} \leq C\varepsilon_0 \quad \text{for } x \in \mathbb{R}^n.$$

For $C\varepsilon_0 \leq \varepsilon_1$, we get from the Gap Lemma 3.4.1 that f parametrises planes and round spheres. In case it parametrises at least one round sphere, we know that $\int_{\Sigma^\infty} |A_{f^\infty}|^2 \,\mathrm{d}\mu_{f^\infty} \geq 8\pi$, which is impossible for ε_0 small enough. Therefore f parametrises planes or even $\Sigma^\infty = \emptyset$, that is f^∞ is trivial.

To get non-triviality, we consider

$$f^j := r_{t_j}^{-1}(f_{t_j + c_0 r_{t_j}^4} - y_{t_j}) \to f^\infty \quad \text{smoothly on compact subsets of } \mathbb{R}^n$$

with $y_{t_j} := x_{\hat{t}_j}$, $\hat{t}_j := t_j + c_0 r_{t_j}^4$ by (3.3.5), that is

$$\int_{\overline{B_{r_{\hat{t}_j}/r_{t_j}}(0)}} |A_{f^j}|^2 \,\mathrm{d}\mu_{f^j} = \int_{\overline{B_{r_{\hat{t}_j}}(x_{\hat{t}_j})}} |A_{f_{\hat{t}_j}}|^2 \,\mathrm{d}\mu_{f_{\hat{t}_j}} \geq \varepsilon_0.$$

Clearly, if we show

$$\liminf_{t\to T} \frac{r_{t+c_0 r_t^4}}{r_t} < \infty, \tag{3.A.1}$$

we can select a subsequence $t_j \to T$ with $r_{\hat{t}_j}/r_{t_j} < \varrho < \infty$ and see from above by smooth convergence in compact subsets that

$$\int_{B_\varrho(0)} |A_{f^\infty}|^2 \,\mathrm{d}\mu_{f^\infty} \geq \varepsilon_0,$$

hence $A_{f^\infty} \not\equiv 0$, and f^∞ does not parametrise a union of planes, in particular $\Sigma^\infty \neq \emptyset$.

Indeed for $\liminf_{t\to T} r_{t+c_0 r_t^4}/r_t > \Gamma > 0$, we can choose $t_0 < T$ such that $r_{t+c_0 r_t^4} \geq \Gamma r_t \; \forall t_0 \leq t < T$. Putting $t_{j+1} := t_j + c_0 r_{t_j}^4$, we see

$$r_{t_j} \geq r_{t_0}\Gamma^j > 0 \quad \forall j \in \mathbb{N}_0.$$

Clearly for ε small and using (2.1.21)

$$r_t \leq \operatorname{diam}(f_t(\Sigma)) \leq C\mu_{f_t}(\Sigma)^{1/2}\mathcal{W}(f_t)^{1/2} \leq C\mu_{f_t}(\Sigma)^{1/2},$$

as $\mathcal{W}(f_t) \leq \mathcal{W}(f_0)$. The first variation formula (1.3.3) for the areas yields

$$\frac{d}{dt}\mu_{f_t}(\Sigma) = -\int_\Sigma \langle \vec{\mathbf{H}}_{f_t}, \partial_t f_t\rangle \, d\mu_{f_t} \leq C\Big(\int_\Sigma |\partial_t f|^2 \, d\mu_{f_t}\Big)^{1/2},$$

and, as f evolves as gradient flow of the Willmore functional,

$$\mu_{f_t}(\Sigma) \leq \mu_{f_0}(\Sigma) + Ct^{1/2}\Big(\int_0^T\int_\Sigma |\partial_t f|^2 \, d\mu_{f_t}\, dt\Big)^{1/2} \leq C(1+t^{1/2}).$$

Hence

$$\begin{aligned} r_t^4 &\leq C(1+t),\\ 1+t_{j+1} &= 1+t_j+c_0 r_{t_j}^4 \leq (1+Cc_0)(1+t_j), \end{aligned}$$

and

$$0 < r_{t_0}^4\Gamma^{4j} \leq r_{t_j}^4 \leq C(1+t_j) \leq C(1+Cc_0)^j(1+t_0),$$

which yields $\Gamma \leq (1+Cc_0)^{1/4}$ for $j \to \infty$ and (3.A.1). □

References

[Bl09] S. Blatt, *A singular example for the Willmore flow*, Analysis, **29** (2009), 407–440.

[ChFaSch09] R. Chill, E. Fasangová and R. Schätzle, *Willmore blow ups are never compact*, Duke Mathematical Journal, **147** (2009), 345–376.

[dC] M. P. do Carmo, "Riemannian Geometry", Birkhäuser, Boston - Basel - Berlin, 1992.

[Ei] S. D. Eidel'man, "Parabolic Systems", North-Holland, Amsterdam, 1969.

[Ha82] R. HAMILTON, *Three-manifolds with positive Ricci curvature*, Journal of Differential Geometry, **17** (1982), 255–306.

[KuSch01] E. KUWERT and R. SCHÄTZLE, *The Willmore Flow with small initial energy*, Journal of Differential Geometry, **57** (2001), 409–441.

[KuSch02] E. KUWERT and R. SCHÄTZLE, *Gradient flow for the Willmore functional*, Communications in Analysis and Geometry, **10** (2002), 307–339.

[MaSi02] U. F. MAYER and G. SIMONETT, *A numerical scheme for radially symmetric solutions of curvature driven free boundary problems, with applications to the Willmore Flow*, Interfaces and Free Boundaries, **4** (2002), 89–109.

[Ms77] W. S. MASSEY, "Algebraic Topology: An Introduction", Springer Verlag, New York- Heidelberg - Berlin, 1977.

[MiSi73] J. H. MICHAEL and L. SIMON, *Sobolev and mean-value inequalities on generalized submanifolds of* $\mathbb{R}^n$, Communications on Pure and Applied Mathematics, **26** (1973), 361–379.

4 Conformal parametrization

4.1 Conformal parametrization

By a theorem of Huber, see [Hu57], a complete, oriented surface Σ immersed in $\mathbb{R}^n$ with $\int_\Sigma |A_\Sigma|^2 \, d\mathcal{H}^2 < \infty$ is conformally equivalent to a compact Riemann surface with finitely many points deleted. Moreover if M is simply connected and non-compact, it is conformally equivalent to $\mathbb{C} = \mathbb{R}^2$, that is there exists

$$f : \mathbb{R}^2 \xrightarrow{\approx} \Sigma \hookrightarrow \mathbb{R}^n$$

with conformal pull-back metric $f^* g_{\text{euc}} = e^{2u} g_{\text{euc}}$. The aim of this talk is to study the behaviour of the conformal exponent u at infinity following the work of Müller and Šverák in [MuSv95]. We will restrict our presentation to $n = 3$.

Theorem 4.1.1 ([MuSv95]). *Let* $f : \mathbb{R}^2 \to \mathbb{R}^3$ *be a smooth complete conformal immersion with pull-back metric* $g = f^* g_{\text{euc}} = e^{2u} g_{\text{euc}}$ *and square integrable second fundamental form satisfying*

$$\int_{\mathbb{R}^2} K_g \, d\mu_g = 0, \tag{4.1.1}$$

$$\int_{\mathbb{R}^2} |K_g| \, d\mu_g \leq 8\pi - \delta \tag{4.1.2}$$

for some $\delta > 0$. *Then*

$$-\Delta_g u = K_g \quad \text{in } \mathbb{R}^2, \tag{4.1.3}$$

$\lambda := \lim_{z \to \infty} u(z) \in \mathbb{R}$ *exists and*

$$\| u - \lambda \|_{L^\infty(\mathbb{R}^2)}, \| Du \|_{L^2(\mathbb{R}^2)}, \| D^2 u \|_{L^1(\mathbb{R}^2)} \leq C(\delta) \int_{\mathbb{R}^2} |A|^2 \, d\mu_g. \tag{4.1.4}$$

Moreover

$$\lim_{y \to \infty} \frac{|f(y)|}{|y|} = e^\lambda. \tag{4.1.5}$$

Proof. (4.1.3) is immediate from elementary differential geometry. By conformal transformation to the Euclidean Laplacian Δ this reads $-\Delta u = K_g e^{2u}$. Recalling for the Gauß map $\nu : \mathbb{R}^2 \to S^2$ being a smooth normal, which is uniquely determined up to the sign and which exists globally, as $\mathbb{R}^2$ is simply connected, that $\nu^* \mathrm{vol}_{S^2} = K_g \mathrm{vol}_g$, we get

$$-\Delta u = *\nu^* \mathrm{vol}_{S^2}.$$

As by (4.1.1)

$$\int_{\mathbb{R}^2} \nu^* \mathrm{vol}_{S^2} = \int_{\mathbb{R}^2} K_g \mathrm{vol}_g = 0$$

and as the Jacobian of ν is $J\nu = |K_g|$ we get from (4.1.2)

$$\int_{\mathbb{R}^2} J_{\mathbb{R}^2} \nu \, \mathrm{d}\mathcal{L}^2 = \int_{\mathbb{R}^2} J\nu \, \mathrm{d}\mu_g = \int_{\mathbb{R}^2} |K_g| \, \mathrm{d}\mu_g < 8\pi - \delta.$$

Together this satisfies the assumptions of the following Proposition 4.1.1 for $\varphi = \nu$, and there is a solution $v : \mathbb{R}^2 \to \mathbb{R}$

$$\begin{aligned} -\Delta v = *\nu^* \mathrm{vol}_{S^2} = K_g e^{2u} \quad &\text{in } \mathbb{R}^2, \\ \lim_{z \to \infty} v(z) = 0, & \end{aligned} \tag{4.1.6}$$

satisfying, when recalling $|D\nu| = |A|$,

$$\begin{aligned} \| v \|_{L^\infty(\mathbb{R}^2)}, \| Dv \|_{L^2(\mathbb{R}^2)}, \| D^2 v \|_{L^1(\mathbb{R}^2)} &\leq C(\delta) \int_{\mathbb{R}^2} |D\nu|^2 \, \mathrm{d}\mu_g \\ &= C(\delta) \int_{\mathbb{R}^2} |A|^2 \, \mathrm{d}\mu_g. \end{aligned} \tag{4.1.7}$$

Clearly $u - v$ is harmonic on the whole of $\mathbb{R}^2$ and the metric $e^{2(u-v)}$ is still complete, since v is bounded. Then by Proposition 4.B.1 the map

$$z \mapsto u(z) - v(z) - \alpha \log |z|$$

is harmonic and bounded at infinity for some $\alpha \in \mathbb{R}$. Therefore it is the real part of a holomorphic function in $\mathbb{R}^2 - \{0\}$. As $u - v$ is already harmonic on $\mathbb{R}^2$, it is the real part of a holomorphic function in $\mathbb{R}^2$, and therefore $\alpha \log |.|$ is the real part of a holomorphic function in $\mathbb{R}^2 - \{0\}$. This is only possible, if $\alpha = 0$. Then $u - v$ is bounded, hence $u - v \equiv \lambda$ is constant, and (4.1.4) follows.

To prove (4.1.5), we put $f_\varepsilon := \varepsilon(f(./\varepsilon) - f(0)) : \mathbb{R}^2 \to \mathbb{R}^3$. Clearly $g_\varepsilon := f_\varepsilon^* g_{\mathrm{euc}} = g(./\varepsilon) = e^{2u_\varepsilon} g_{\mathrm{euc}}$ with $u_\varepsilon = u(./\varepsilon)$ and for $\varrho > 0$

$$\int_{\mathbb{R}^2 - B_\varrho(0)} |Du_\varepsilon|^2 \, \mathrm{d}\mathcal{L}^2 = \int_{\mathbb{R}^2 - B_{\varrho/\varepsilon}(0)} |Du|^2 \, \mathrm{d}\mathcal{L}^2 \to 0 \quad \text{for } \varepsilon \to 0$$

and

$$u_\varepsilon \to \lambda \quad \text{uniformly on } \mathbb{R}^2 - B_\varrho(0). \tag{4.1.8}$$

Likewise the Gauß map ν_ε of f_ε is given $\nu_\varepsilon(x) := \nu(x/\varepsilon)$ and for $\varrho > 0$

$$\begin{aligned}\int\limits_{\mathbb{R}^2 - B_\varrho(0)} |D\nu_\varepsilon|^2 \, d\mathcal{L}^2 &= \int\limits_{\mathbb{R}^2 - B_\varrho(0)} |D\nu_\varepsilon|^2_{g_\varepsilon} \, d\mu_{g_\varepsilon} \\ &= \int\limits_{\mathbb{R}^2 - B_{\varrho/\varepsilon}(0)} |D\nu|^2_g \, d\mu_g \\ &= \int\limits_{\mathbb{R}^2 - B_{\varrho/\varepsilon}(0)} |A|^2 \, d\mu_g \to 0 \quad \text{for } \varepsilon \to 0.\end{aligned}$$

Next by (4.1.4) and $|\partial f| = e^u$, we get $|Df| \le C$ for some $C < \infty$ and $\mathrm{lip}_{\mathbb{R}^2} f \le C$, hence

$$|f_\varepsilon(x)| \le C|x|, \quad |Df_\varepsilon| \le C.$$

Since f_ε is conformal, we use the formula of Proposition 4.B.2

$$|D^2 f_\varepsilon|^2 = e^{2u_\varepsilon}\left(4|Du_\varepsilon|^2 + |D\nu_\varepsilon|^2\right)$$

and get for $\varrho > 0$ recalling that u, u_ε are uniformly bounded

$$\int\limits_{\mathbb{R}^2 - B_\varrho(0)} |D^2 f_\varepsilon|^2 \, d\mathcal{L}^2 \le C \int\limits_{\mathbb{R}^2 - B_\varrho(0)} \left(|Du_\varepsilon|^2 + |D\nu_\varepsilon|^2\right) d\mathcal{L}^2 \to 0 \quad \text{for } \varepsilon \to 0,$$

$$\begin{aligned}\int\limits_{\mathbb{R}^2} |D^2 f_\varepsilon|^2 \, d\mathcal{L}^2 &\le C \int\limits_{\mathbb{R}^2} \left(|Du_\varepsilon|^2 + |D\nu_\varepsilon|^2\right) d\mathcal{L}^2 \\ &= C \int\limits_{\mathbb{R}^2} \left(|Du|^2 + |D\nu|^2\right) d\mathcal{L}^2 \\ &= C \int\limits_{\mathbb{R}^2} |Du|^2 \, d\mathcal{L}^2 + C \int\limits_{\mathbb{R}^2} |A|^2 \, d\mu_g < \infty.\end{aligned}$$

Then for any ε_k there is a subsequence such that

$$f_{\varepsilon_k} \to f_0 \quad \text{weakly in } W^{2,2}(\mathbb{R}^2), \text{ uniformly on compact subsets of } \mathbb{R}^2.$$

f_0 is a conformal immersion with $f_0^* g_{euc} = e^{2\lambda} g_{euc}$ by (4.1.8). Moreover from above for $\varrho > 0$

$$\int\limits_{\mathbb{R}^2 - B_\varrho(0)} |D^2 f_0|^2 \, d\mathcal{L}^2 \leq \lim_{\varepsilon \to 0} \int\limits_{\mathbb{R}^2 - B_\varrho(0)} |D^2 f_\varepsilon|^2 \, d\mathcal{L}^2 = 0,$$

hence $D^2 f_0 = 0$ and $Df_0 \equiv (a, b)$, with $a \perp b, |a| = |b| = e^\lambda$. This yields by $f_0(0) \leftarrow f_\varepsilon(0) = 0$ that

$$|f_0(y)| = |(a, b)y| = e^\lambda |y|,$$

hence, as this is independent of the subsequence,

$$|f_\varepsilon| \to e^\lambda |.| \quad \text{uniformly on compact subsets of } \mathbb{R}^2.$$

We see for $\varepsilon = 1/|y|$ that

$$\frac{f(y)}{|y|} = f_{1/|y|}(y/|y|) + \frac{f(0)}{|y|}$$

and conculde

$$\lim_{y \to \infty} \frac{|f(y)|}{|y|} = \lim_{|y| \to \infty} \left| f_{1/|y|}(y/|y|) + \frac{f(0)}{|y|} \right| = e^\lambda,$$

which is (4.1.5), and the theorem is proved. □

Proposition 4.1.1. *Let* $\varphi \in W^{1,2}_{loc}(\mathbb{R}^2, S^2)$ *with* $D\varphi \in L^2(\mathbb{R}^2)$ *satisfy*

$$\begin{aligned} \int_{\mathbb{R}^2} J\varphi \, d\mathcal{L}^2 &< 8\pi - \delta, \\ \left| \int_{\mathbb{R}^2} \varphi^* \mathrm{vol}_{S^2} \right| &< 4\pi. \end{aligned} \tag{4.1.9}$$

Then there exists a unique, bounded solution $v : \mathbb{R}^2 \to \mathbb{R}$

$$\begin{aligned} -\Delta v &= *\varphi^* \mathrm{vol}_{S^2} \quad \text{in } \mathbb{R}^2, \\ \lim_{z \to \infty} v(z) &= 0, \end{aligned} \tag{4.1.10}$$

satisfying

$$\| v \|_{L^\infty(\mathbb{R}^2)}, \| Dv \|_{L^2(\mathbb{R}^2)}, \| D^2 v \|_{L^1(\mathbb{R}^2)} \leq C\delta^{-2} \int\limits_{\mathbb{R}^2} |D\varphi|^2 \, d\mathcal{L}^2. \tag{4.1.11}$$

Proof. We follow [MuSv95, Corollary 3.3.2]. By [MuSv95, Proposition 3.4.1] using [ScUh83, Section 4], we approximate φ in $W^{1,2}_{\text{loc}}(\mathbb{R}^2, S^2)$ with $D\varphi \in L^2(\mathbb{R}^2)$ by a smooth map which is constant outside a compact subset of $\mathbb{R}^2$ while keeping (4.1.9). and we will first assume that φ is smooth and constant outside a compact subset of $\mathbb{R}^2$.

Composing φ with the stereographic projection and extending to S^2, the mapping degree as map $S^2 \to S^2$ satsifies

$$\int\limits_{\mathbb{R}^2} \varphi^* \mathrm{vol}_{S^2} = 4\pi \deg(\varphi) \in 4\pi\mathbb{Z},$$

hence by (4.1.9)

$$\deg(\varphi) = 0.$$

Counting preimages for regular values according to their orientation, we get by Sard's theorem

$$\mathcal{H}^0(\varphi^{-1}(p)) \text{ is even for almost all } p \in S^2,$$

hence again by (4.1.9)

$$\mathrm{vol}_{S^2}(\varphi(\mathbb{R}^2)) \le \frac{1}{2} \int\limits_{\mathbb{R}^2} J\varphi \, \mathrm{d}\mathcal{L}^2 < 4\pi - \delta/2,$$

in particular $\mathrm{vol}_{S^2}(S^2 - \varphi(\mathbb{R}^2)) > \delta/2$. As $\varphi(\mathbb{R}^2)$ is a Borel set, actually compact, in particular measurable, we can choose a compact subset $K \subseteq S^2 - \varphi(\mathbb{R}^2)$ with

$$\mathrm{vol}_{S^2}(K) > \delta/2 \tag{4.1.12}$$

and put $U := S^2 - K \supseteq \varphi(\mathbb{R}^2)$ open.

Now we continue with the construction of [MuSv95, Corollary 3.3.2]. We consider the map $T_0 : S^2 - \{(0,0,-1)\} \xrightarrow{\approx} B_2(0) \subseteq \mathbb{R}^2$ defined by

$$\begin{array}{ccc} & & S^2 - \{(0,0,-1)\} \\ & \overset{\phi}{\nearrow} & \\ [0,\pi[\times\mathbb{R} & & \approx \Big\downarrow T_0 \\ & \underset{\psi}{\searrow} & \\ & & B_2(0), \end{array}$$

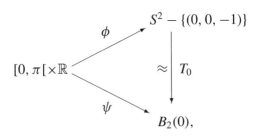

in polar coordinates

$$\phi(r,\theta) := (\sin r\cos\theta, \sin r\sin\theta, \cos r),$$
$$\psi(r,\theta) := \varrho(r)(\cos\theta, \sin\theta).$$

We calculate

$$D\phi = \begin{pmatrix} \cos r\cos\theta & -\sin r\sin\theta \\ \cos r\sin\theta & \sin r\cos\theta \\ -\sin r & 0 \end{pmatrix},$$

$$(J\phi)^2 = \det(D\phi^T D\phi) = Gram(\partial_r\phi, \partial_\theta\phi) = |\partial_r\phi|^2\,|\partial_\theta\phi|^2 = \sin^2 r,$$

and

$$D\psi = \begin{pmatrix} \varrho'(r)\cos\theta & -\varrho(r)\sin\theta \\ \varrho'(r)\sin\theta & \varrho(r)\cos\theta \end{pmatrix},$$

$$J\psi = |\det(D\psi)| = |\varrho\varrho'(r)|.$$

We seek $J\phi = J\psi$ that is $|(\frac{1}{2}\varrho^2)'(r)| = \sin r$, hence $\varrho(r)^2 = 2(c \mp \cos r)$ and $\varrho(r) = \sqrt{2(c\mp\cos r)}$. Putting $\varrho(r) = \sqrt{2(1-\cos r)} = 2\sin(r/2)$, we get $J\phi = J\psi$. Clearly with this definition, T_0 is bijective and smooth on $S^2 - \{(0,0,\pm 1)\}$. More precisely for $(x,y,z) = (\sin r\cos\theta, \sin r\sin\theta, \cos r) \in S^2 - \{(0,0,-1)\}$, we see

$$T_0(x,y,z) = \varrho(r)(\cos\theta,\sin\theta) = \frac{2\sin(r/2)}{\sin r}(\sin r\cos\theta, \sin r\sin\theta)$$
$$= \frac{2\sin(r/2)}{\sin r}(x,y).$$

Recalling $\sin r = 2\sin(r/2)\cos(r/2), 2\cos(r/2)^2 = 1+\cos r = 1+z$, we obtain

$$\frac{2\sin(r/2)}{\sin r} = \frac{2\sin(r/2)}{2\sin(r/2)\cos(r/2)} = \frac{1}{\cos(r/2)} = \sqrt{\frac{2}{1+z}}$$

and

$$T_0(x,y,z) = \sqrt{\frac{2}{1+z}}(x,y).$$

Therefore T_0 is smooth on the whole of $S^2 - \{(0,0,-1)\}$ and, since $J\phi = J\psi$ and both ϕ and ψ are orientation preserving, we get

$$\phi^*\mathrm{vol}_{S^2} = \psi^* dx_1 \wedge dx_2 = \phi^* T_0^* dx_1 \wedge dx_2$$

and

$$T_0^* dx_1 \wedge dx_2 = \mathrm{vol}_{S^2}\lfloor(S^2 - \{(0,0,-1)\}$$

that is T_0 is volume preserving.

Moreover

$$|\partial_x T_0|, |\partial_y T_0| = \sqrt{\frac{2}{1+z}}, \quad |\partial_z T_0| = \frac{|(x, y)|}{\sqrt{2}(1+z)^{3/2}}.$$

We observe

$$|(x, y)|^2 = x^2 + y^2 = 1 - z^2 = (1+z)(1-z)$$

and for the tangential projection

$$|e_3^{\tan}| = \sqrt{1 - |e_3^{\perp}|^2} = \sqrt{1 - z^2} = |(x, y)|,$$

hence for the gradient on S^2

$$|\nabla_z^{S^2} T_0| = |e_3 \cdot \nabla^{S^2} T_0| = |e_3^{\tan}|\, |\partial_z T_0| = \frac{1-z}{\sqrt{2(1+z)}} \leq \sqrt{\frac{2}{1+z}}.$$

Next as

$$|(x, y, z) - (0, 0, -1)|^2 = x^2 + y^2 + (1+z)^2 = 1 - z^2 + (1+z)^2 = 2(1+z),$$

we obtain

$$|\nabla^{S^2} T_0(x, y, z)| \leq C / |(x, y, z) - (0, 0, -1)|.$$

Now we put $\nu_0 := (0, 0, -1)$, $O_{\nu_0} = \mathrm{id}_{S^2}$ and define for $\nu \neq \nu_0$ the orthogonal transformation

$$O_\nu x := x - 2\frac{(\nu_0 - \nu)(\nu_0 - \nu)^T}{|\nu_0 - \nu|^2}.$$

Clearly $O_\nu \nu_0 = \nu$, $O_\nu \nu = \nu_0$ and the dependence $\nu \to O_\nu$ is smooth in $S^2 - \{\nu_0\}$. We put $T_\nu := T_0 \circ O_\nu : S^2 - \{\nu\} \xrightarrow{\approx} B_2(0)$ and see form above

$$\begin{aligned} T_\nu^* dx_1 \wedge dx_2 &= O_\nu^* T_0^* dx_1 \wedge dx_2 = O_\nu^* \Big(\mathrm{vol}_{S^2} \lfloor (S^2 - \{\nu_0\}) \Big) \\ &= \mathrm{vol}_{S^2} \lfloor (S^2 - \{\nu\}), \end{aligned} \tag{4.1.13}$$

$$|\nabla^{S^2} T_\nu(p)| = |\nabla^{S^2} T_0(O_\nu p)| \leq C / |O_\nu p - \nu_0| = C / |p - \nu|.$$

Clearly $K \subseteq S^2 - \varphi(\mathbb{R}^2) \neq S^2$, and we may assume $\nu_0 \notin K$. Then we define $T : U \to B_2(0)$ by

$$T(p) := \frac{1}{\mathrm{vol}_{S^2}(K)} \int_K T_\nu(p) \,\mathrm{d}\,\mathrm{vol}_{S^2}(\nu).$$

Since $d(\nu, K) > 0$ for all $\nu \in U$ and by the continuous, hence measurable dependence $\nu \mapsto T_\nu$ this is well defined and by (4.1.13) T is smooth,

$$T^* dx_1 \wedge dx_2 = \mathrm{vol}^2_S | U$$

and

$$\begin{aligned} |\nabla^{S^2} T(p)| &\le \frac{1}{\mathrm{vol}_{S^2}(K)} \int_K |\nabla^{S^2} T_\nu(p)| \, \mathrm{d}\,\mathrm{vol}_{S^2}(\nu) \\ &\le C\delta^{-1} \int_{S^2} \frac{1}{|p-\nu|} \, \mathrm{d}\,\mathrm{vol}_{S^2}(\nu) \le C\delta^{-1}. \end{aligned}$$

This yields for $T \circ \varphi : \mathbb{R}^2 \to \mathbb{R}^2$

$$\varphi^* \mathrm{vol}_{S^2} = (T \circ \varphi)^* dx_1 \wedge dx_2 = \det D(T \circ \varphi) dx_1 \wedge dx_2$$

and by Theorem 4.A.2

$$\begin{aligned} \| *\varphi^* \mathrm{vol}_{S^2} \|_{\mathcal{H}^1(\mathbb{R}^2)} &= \| \det D(T \circ \varphi) \|_{\mathcal{H}^1(\mathbb{R}^2)} \le C \| D(T \circ \varphi) \|^2_{L^2(\mathbb{R}^2)} \\ &\le C\delta^{-2} \| D\varphi \|^2_{L^2(\mathbb{R}^2)} . \end{aligned}$$

Then by Theorem 4.A.1 or by Theorem 4.A.3 apart form the $W^{2,1}$-estimates there exists a unique v

$$\begin{aligned} -\Delta v = *\varphi^* \mathrm{vol}_{S^2} = \det(D(T \circ \varphi)) \quad &\text{in } \mathbb{R}^2, \\ \lim_{z \to \infty} v(z) = 0. & \end{aligned}$$

v is continuous and satisfies the estimates

$$\begin{aligned} \| v \|_{L^\infty(\mathbb{R}^2)}, \| Dv \|_{L^2(\mathbb{R}^2)}, \| D^2 v \|_{L^1(\mathbb{R}^2)} &\le C \| *\varphi^* \mathrm{vol}_{S^2} \|_{\mathcal{H}^1(\mathbb{R}^2)} \\ &\le C\delta^{-2} \| D\varphi \|^2_{L^2(\mathbb{R}^2)}, \end{aligned}$$

and the proposition is proved for smooth φ which is constant outside a compact set.

For $\varphi \in W^{1,2}_{\mathrm{loc}}(\mathbb{R}^2, S^2)$ with $D\varphi \in L^2(\mathbb{R}^2)$, we approximate φ with [MuSv95] Proposition 3.4.1 using [ScUh83, Section 4] by smooth maps $\varphi_m : \mathbb{R}^2 \to S^2$ which are constant outside a compact set and

$$\begin{aligned} \varphi_m \to \varphi \quad &\text{strongly in } L^2_{\mathrm{loc}}(\mathbb{R}^2), \\ D\varphi_m \to D\varphi \quad &\text{strongly in } L^2(\mathbb{R}^2), \\ \varphi_m, D\varphi_m \to \varphi, D\varphi \quad &\text{pointwise almost everywhere in } \mathbb{R}^2. \end{aligned}$$

Clearly φ_m satisfy (4.1.9) for m large. The above construction yields smooth maps $\Phi_m := T_m \circ \varphi_m : \mathbb{R}^2 \to B_2(0) \subseteq \mathbb{R}^2$ for m large with

$$\begin{aligned} *\varphi_m^* \mathrm{vol}_{S^2} = \det D\Phi_m \quad &\text{in } \mathbb{R}^2, \\ \| D\Phi_m \|_{L^2(\mathbb{R}^2)} \le C\delta^{-1} \| D\varphi_m \|_{L^2(\mathbb{R}^2)} . & \end{aligned}$$

Clearly

$$\begin{aligned}\limsup_{m\to\infty} \| D\Phi_m \|_{L^2(\mathbb{R}^2)} &\le \limsup_{m\to\infty} C\delta^{-1} \| D\varphi_m \|_{L^2(\mathbb{R}^2)} \\ &= C\delta^{-1} \| D\varphi \|_{L^2(\mathbb{R}^2)} < \infty,\end{aligned}$$

hence for a subsequence

$$\begin{aligned}\Phi_m &\to \Phi \quad \text{strongly in } L^2_{\text{loc}}(\mathbb{R}^2), \\ D\Phi_m &\to D\Phi \quad \text{weakly in } L^2(\mathbb{R}^2),\end{aligned}$$

with $\Phi \in W^{1,2}_{\text{loc}}(\mathbb{R}^2, \mathbb{R}^2)$ and

$$\| D\Phi \|_{L^2(\mathbb{R}^2)} \le C\delta^{-1} \| D\varphi \|_{L^2(\mathbb{R}^2)} .$$

We claim the following convergences

$$\begin{aligned} *\varphi_m^*\text{vol}_{S^2} &\to *\varphi^*\text{vol}_{S^2} \quad \text{strongly in } L^1(\mathbb{R}^2), \\ \int_{\mathbb{R}^2} \det(D\Phi_m)\lambda \, d\mathcal{L}^2 &\to \int_{\mathbb{R}^2} \det(D\Phi)\lambda \, d\mathcal{L}^2 \text{ for all } \lambda \in C_0^\infty(\mathbb{R}^2).\end{aligned} \tag{4.1.14}$$

Recalling $\text{vol}_{S^2} = (dx_1 \wedge dx_2 \wedge dx_3) \llcorner \nu_{S^2}$, where ν_{S^2} denotes the outer unit normal at φ, we see

$$*\varphi^*\text{vol}_{S^2} = \det(\varphi, \partial_1\varphi, \partial_2\varphi) \leftarrow \det(\varphi_m, \partial_1\varphi_m, \partial_2\varphi_m) = *\varphi_m^*\text{vol}_{S^2}$$

pointwise almost everywhere on $\mathbb{R}^2$. Moreover

$$| * \varphi_m^*\text{vol}_{S^2}| = |\det(\varphi_m, \partial_1\varphi_m, \partial_2\varphi_m)| \le |D\varphi_m|^2$$

and, as $D\varphi_m \to D\varphi$ strongly in $L^2(\mathbb{R}^2)$, we get by Vitali's convergence theorem the first convergence in (4.1.14).

Next we calculate for $\lambda \in C_0^\infty(\mathbb{R}^2)$ that

$$\begin{aligned}\int_{\mathbb{R}^2} \det(D\Phi)\lambda \, d\mathcal{L}^2 &= \int_{\mathbb{R}^2} \left(\partial_1\Phi^1\partial_2\Phi^2 - \partial_2\Phi^1\partial_1\Phi^2\right)\lambda \, d\mathcal{L}^2 \\ &= \int_{\mathbb{R}^2} \left(\partial_1(\Phi^1\lambda)\partial_2\Phi^2 - \partial_2(\Phi^1\lambda)\partial_1\Phi^2\right) d\mathcal{L}^2 \\ &\quad - \int_{\mathbb{R}^2} \left(\Phi^1\partial_2\Phi^2\partial_1\lambda - \Phi^1\partial_1\Phi^2\partial_2\lambda\right) d\mathcal{L}^2.\end{aligned}$$

Now $\Phi^1\lambda, \Phi^2 \in W^{1,2}_{loc}(\mathbb{R}^2)$ and $\Phi^1\lambda$ has compact support and by approximation by functions of $C_0^\infty(\mathbb{R}^2)$, we see that the first term on the ride hand side vanishes. Therefore

$$\int_{\mathbb{R}^2} \det(D\Phi)\lambda \, d\mathcal{L}^2 = -\int_{\mathbb{R}^2} \left(\Phi^1\partial_2\Phi^2\partial_1\lambda - \Phi^1\partial_1\Phi^2\partial_2\lambda\right) d\mathcal{L}^2$$

and the same is true for Φ replaced by Φ_m. Then by $\Phi_m \to \Phi$ strongly in $L^2_{loc}(\mathbb{R}^2)$, $D\Phi_m \to D\Phi$ weakly in $L^2(\mathbb{R}^2)$ and $D\lambda \in C_0^0(\mathbb{R}^2)$

$$\begin{aligned}\int_{\mathbb{R}^2} \det(D\Phi_m)\lambda \, d\mathcal{L}^2 &= -\int_{\mathbb{R}^2} \left(\Phi^1_m\partial_2\Phi^2_m\partial_1\lambda - \Phi^1_m\partial_1\Phi^2_m\partial_2\lambda\right) d\mathcal{L}^2 \\ &\to -\int_{\mathbb{R}^2} \left(\Phi^1\partial_2\Phi^2\partial_1\lambda - \Phi^1\partial_1\Phi^2\partial_2\lambda\right) d\mathcal{L}^2 \\ &= \int_{\mathbb{R}^2} \det(D\Phi)\lambda \, d\mathcal{L}^2,\end{aligned}$$

which is the second convergence in (4.1.14).

(4.1.14) implies that

$$*\varphi^*\mathrm{vol}_{S^2} = \det D\Phi \quad \text{in } \mathbb{R}^2. \tag{4.1.15}$$

As above by Theorem 4.A.2

$$\begin{aligned}\| *\varphi^*\mathrm{vol}_{S^2} \|_{\mathcal{H}^1(\mathbb{R}^2)} &= \| \det D\Phi \|_{\mathcal{H}^1(\mathbb{R}^2)} \\ &\le C \| D\Phi \|^2_{L^2(\mathbb{R}^2)} \le C\delta^{-2} \| D\varphi \|^2_{L^2(\mathbb{R}^2)}\end{aligned}$$

and by Theorem 4.A.1 or by Theorem 4.A.3 apart form the $W^{2,1}$-estimates there exists a unique v

$$\begin{aligned}-\Delta v = \det(D\Phi) &= *\varphi^*\mathrm{vol}_{S^2} \quad \text{in } \mathbb{R}^2, \\ \lim_{z\to\infty} v(z) &= 0.\end{aligned}$$

v is continuous and satisfies the estimates

$$\begin{aligned}\| v \|_{L^\infty(\mathbb{R}^2)}, \| Dv \|_{L^2(\mathbb{R}^2)}, \| D^2 v \|_{L^1(\mathbb{R}^2)} &\le C \| *\varphi^*\mathrm{vol}_{S^2} \|_{\mathcal{H}^1(\mathbb{R}^2)} \\ &\le C\delta^{-2} \| D\varphi \|^2_{L^2(\mathbb{R}^2)},\end{aligned}$$

and the proposition is proved in full generality. □

Appendix

4.A $\mathcal{H}^1$-Estimates

We recall the following results concerning the Hardy space $\mathcal{H}^1$.

Theorem 4.A.1 ([FSt72], [MuSv95, Theorem 3.2.1]). *For $\alpha \in \mathcal{H}^1(\mathbb{R}^2)$ there existis a unique $v \in W^{2,1}_{\text{loc}}(\mathbb{R}^2)$ of*

$$-\Delta v = \alpha \quad \textit{in } \mathbb{R}^2,$$
$$\lim_{z\to\infty} v(z) = 0.$$

v is continuous and satisfies the estimates

$$\| v \|_{L^\infty(\mathbb{R}^2)}, \| Dv \|_{L^2(\mathbb{R}^2)}, \| D^2 v \|_{L^1(\mathbb{R}^2)} \leq C \| \alpha \|_{\mathcal{H}^1(\mathbb{R}^2)} .$$

Theorem 4.A.2 ([CLMSe93]). *For $\varphi \in W^{1,n}_{\text{loc}}(\mathbb{R}^n, \mathbb{R})$ with $D\varphi \in L^n(\mathbb{R}^n, \mathbb{R}^n)$, the determinant $\det(D\varphi)$ belongs to $\mathcal{H}^1(\mathbb{R}^n)$ and satisfies*

$$\| \det(D\varphi) \|_{\mathcal{H}^1(\mathbb{R}^n)} \leq C_n \| D\varphi \|^n_{L^n(\mathbb{R}^2)} .$$

Remark 4.A.1. These results admit a solution to the equation $\Delta v = \det(D\varphi)$ for $\varphi \in W^{1,2}(\mathbb{R}^2, \mathbb{R}^2)$ which is continuous and satisfies the estimates above. This was earlier proved by Wente apart from the $W^{2,1}$-estimate, but which would suffice for our purposes.

Theorem 4.A.3 ([We69]). *For $\varphi \in W^{1,2}(\mathbb{R}^2, \mathbb{R}^2)$ there exists a unique solution $v \in W^{1,2}(\mathbb{R}^2) \cap C^0(\mathbb{R}^2)$ of*

$$-\Delta v = \det(D\varphi) \quad \textit{in } \mathbb{R}^2,$$
$$\lim_{z\to\infty} v(z) = 0.$$

v satisfies the estimates

$$\| v \|_{L^\infty(\mathbb{R}^2)}, \| Dv \|_{L^2(\mathbb{R}^2)} \leq C \| D\varphi \|^2_{L^2(\mathbb{R}^2)} .$$

4.B Auxiliary results

Proposition 4.B.1 ([MuSv95] Lemma 4.1.2). *Let $u : \mathbb{C} - B_1(0) \to \mathbb{R}$ be a harmonic function such that the metric $e^{2u}\delta_{ij}$ is complete at infinity. Then for some $\alpha \geq -1$ the map*

$$z \mapsto u(z) - \alpha \log |z|$$

is harmonic and bounded at infinity.

Proof. We follow [MuSv95, Lemma 4.1.2] with a simplfied argument from [Os69, Lemma 9.6]. Putting

$$\alpha := \frac{1}{2\pi} \int_0^{2\pi} \langle \nabla u(r\cos\varphi, r\sin\varphi), (r\cos\varphi, r\sin\varphi)\rangle \, d\varphi \in \mathbb{R} \quad \text{for } r > 1,$$

we see that $u - \alpha \log|.|$ is the real part of a holomorphic function Φ in $\mathbb{C} - B_1(0)$. We do a Laurent expansion of Φ in $\mathbb{C} - B_1(0)$ in the form

$$\Phi(z) = \sum_{k=-\infty}^{\infty} c_k z^k \quad \text{for } |z| > 1$$

and suitable $c_k \in \mathbb{C}$. Putting

$$\Phi_0(z) := \sum_{k=-\infty}^{-1} c_k z^k \quad \text{for } |z| > 1,$$
$$\Psi(z) := \sum_{k=0}^{\infty} c_k z^k \quad \text{for } z \in \mathbb{C},$$

where we observe that the power series in the second definition converges for $|z| > 1$, hence on all of $\mathbb{C}$ and defines an entire function. Clearly $\Phi = \Phi_0 + \Psi$, and $\Phi_0(z) \to 0$ for $z \to \infty$, hence Φ_0 is bounded at infinity. Then for any integer $N \geq \alpha$

$$e^{u(z)} = |z|^\alpha e^{Re\Phi(z)} \leq C|z|^N e^{Re\Psi(z)} \quad \text{for } |z| > 2,$$

and the metric

$$|z^N e^{\Psi(z)}|^2 \delta_{ij} \text{ is complete at infinity} \tag{4.B.1}$$

too. As $\mathbb{C}$ is simply connected, there exists an entire function F with $F'(z) = z^N e^{\psi(z)}$, $F(0) = 0$. We see

$$F(z) = e^{\Psi(0)} z^{N+1}/(N+1) + z^{N+2} \ldots = z^{N+1} \psi(z)$$

for some entire ψ with $\psi(0) = e^{\Psi(0)}/(N+1) \neq 0$, and there exists locally $\tilde{\lambda}$ holomorphic in a neighbourhood of 0 with $\psi = \tilde{\lambda}^{N+1}, \tilde{\lambda}(0) \neq 0$. Putting $\lambda(z) := z\tilde{\lambda}(z)$, we see that $F = \lambda^{N+1}$ in a neighbourhood of 0 and $\lambda(0) = 0, \lambda'(0) \neq 0$. Then λ admits an inverse g such that $\lambda(g(w)) = w$ in a neighbourhood of 0, in particular

$$g'(0) = \lambda'(0)^{-1} \neq 0. \tag{4.B.2}$$

We extend g holomorphically to the maximal disc $B_R(0) \subseteq \mathbb{C}$ with $0 < R \leq \infty$. Then $F \circ g$ is holomorphic in $B_R(0)$, as F is entire, and in a neighbourhood of 0 we have $F(g(w)) = \lambda^{N+1}(g(w)) = w^{N+1}$, hence

$$F(g(w)) = w^{N+1} \quad \text{for all } w \in B_R(0). \tag{4.B.3}$$

We claim $R = \infty$, that is g extends to an entire function. Indeed, if $R < \infty$, there exists $w_0 \in \partial B_R(0)$ and g cannot be extended to $B_R(0) \cup U(w_0)$ for any neighbourhood $U(w_0)$ of w_0. We consider the path $w(t) = tw_0$ for $0 \leq t < 1$ and its image $\gamma := g \circ w$ under g. By (4.B.3)

$$\begin{aligned} F'(\gamma(t))\gamma'(t) &= \frac{d}{dt} F(g(w(t))) = \frac{d}{dt} w(t)^{N+1} \\ &= \frac{d}{dt}(tw_0)^{N+1} = (N+1) w_0^{N+1} t^N, \end{aligned}$$

hence

$$\begin{aligned} \int_0^1 |\gamma(t)^N e^{\Psi(\gamma(t))}|\, |\gamma'(t)|\, dt &= \int_0^1 |F'(\gamma(t))\gamma'(t)|\, dt \\ &= \int_0^1 (N+1)|w_0|^{N+1} t^N\, dt \\ &= |w_0|^{N+1} = R^{N+1} < \infty. \end{aligned}$$

By (4.B.1), the path γ cannot diverge to infinity, hence $\liminf_{t \nearrow 1} |\gamma(t)| < \infty$, and we select $t_k \nearrow 1$ with $z_k := \gamma(t_k) = g(w(t_k))$ bounded. For a subsequence, we get $z_k \to z_0 \in \mathbb{C}$, $w(t_k) = t_k w_0 \to w_0$ and

$$F(z_0) \leftarrow F(z_k) = F(g(w(t_k))) = w(t_k)^{N+1} \to w_0^{N+1} \neq 0,$$

in particular $z_0 \neq 0$, as $F(0) = 0$, hence $F'(z_0) = z_0^N e^{\psi(z_0)} \neq 0$. Then there exists a local inverse $\tau : U(w_0^{N+1}) \xrightarrow{\approx} U(z_0)$ of F. We choose $\varrho > 0$ small enough such that

$$\forall w \in B_\varrho(w_0) : w^{N+1} \in U(w_0^{N+1})$$

and $w \mapsto \tau(w^{N+1})$ is holomorphic in $B_\varrho(w_0)$. We claim

$$\forall w \in B_\varrho(w_0) \cap B_R(0) : g(w) \in U(z_0) \Longleftrightarrow g(w) = \tau(w^{N+1}).$$

Indeed if $g(w) \in U(z_0)$, then $g(w) = \tau(F(g(w))) = \tau(w^{N+1})$. On the other hand, from the right hand side, we see $g(w) = \tau(w^{N+1}) \in U(z_0)$.

This yields that

$$A := \{w \in B_\varrho(w_0) \cap B_R(0) \mid g(w) = \tau(w^{N+1})\}$$

is open and closed in the convex set $B_\varrho(w_0) \cap B_R(0)$. Recalling $t_k w_0 \to w_0, z_k \to z_0$, we see for k large that $t_k w_0 \in B_\varrho(w_0) \cap B_R(0), g(t_k w_0) = z_k \in U(z_0)$, hence $t_k w_0 \in A$ and A is non-empty. Therefore $A = B_\varrho(w_0) \cap B_R(0)$, as the right hand side is connected, and the holomorphic functions g and $w \mapsto \tau(w^{N+1})$ coincide on $B_\varrho(w_0) \cap B_R(0)$. Putting $g(w) := \tau(w^{N+1})$ for $w \in B_\varrho(w_0)$, this extends g holomorphically to $B_R(0) \cup B_\varrho(w_0)$, which contradicts the definition of R, and hence $R = \infty$, as claimed.

Therefore g extends to an entire function and satisfies (4.B.3) on the whole of $\mathbb{C}$. This yields for $w_1, w_2 \in \mathbb{C}$ that

$$g(w_1) = g(w_2) \Longrightarrow F(g(w_1)) = F(g(w_2)) \Longrightarrow w_1^{N+1} = w_2^{N+1},$$

and g can take any value in $\mathbb{C}$ at most $N+1$-times, hence g must be a polynomial of degree at most $N+1$ by Picard's theorem. As moreover $g(w) = 0$ implies $w^{N+1} = F(g(w)) = F(0) = 0$, hence $w = 0$, we get $g(w) = aw^k$ for some $a \in \mathbb{C}, 0 \leq k \leq N+1$. Then $g'(w) = kaw^{k-1}$ and combining with (4.B.2), we see $k = 1, a \neq 0$ and $g(w) = aw$. Then $w^{N+1} = F(g(w)) = F(aw)$ by (4.B.3) or likewise

$$F(z) = (a^{-1}z)^{N+1} = a^{-N-1}z^{N+1},$$

and F is a polynomial as well. We calculate

$$z^N e^{\Psi(z)} = F'(z) = a^{-N-1}(N+1)z^N$$

and $e^{\Psi(z)} = a^{-N-1}(N+1)$ for $z \neq 0$, hence Ψ is constant.

Therefore

$$z \mapsto u(z) - \alpha \log |z| = Re\Phi(z) = Re\Phi_0(z) + Re\Psi(0)$$

is bounded at infinity, as Φ_0 is bounded at infinity.

Clearly this map is harmonic. Moreover $e^{u(z)} \sim e^{\alpha \log |z|} = |z|^\alpha$ for $z \to \infty$, and, as $e^{2u}\delta_{ij}$ is assumed to be complete, we obtain

$$\infty = \int_2^\infty e^{u(t)}\, dt \leq C \int_2^\infty t^\alpha\, dt$$

and conclude $\alpha \geq -1$. □

Proposition 4.B.2 ([MuSv95, Lemma 4.2.7]). *For a conformal immersion* $f : U \subseteq \mathbb{R}^2 \to \mathbb{R}^3$ *with pull back metric* $g = f^* g_{\text{euc}} = e^{2u} g_{\text{euc}}$ *and Gauß map* $\nu : \mathbb{R}^2 \to S^2$*, we get for the second derivatives*

$$|D^2 f|^2 = e^{2u}\Big(4|Du|^2 + |D\nu|^2\Big)$$

Proof. The equations of Weingarten

$$\partial_{ij} f = \Gamma_{ij}^k \partial_k f + A_{ij}$$

read together with the formulas for the Christoffel symbols in Proposition 1.2.2 (1.2.5)

$$\begin{aligned} T_{11}^1 &= \partial_1 u, \quad T_{12}^1 = T_{21}^1 = \partial_2 u, \ T_{22}^1 = -\partial_1 u, \\ T_{11}^2 &= -\partial_2 u, \ T_{12}^2 = T_{21}^2 = \partial_1 u, \ T_{22}^2 = \partial_2 u, \end{aligned}$$

that

$$\begin{aligned} \partial_{11} f &= \partial_1 u \, \partial_1 f - \partial_2 u \, \partial_2 f + A_{11}, \\ \partial_{12} f &= \partial_2 u \, \partial_1 f + \partial_1 u \, \partial_2 f + A_{12}, \\ \partial_{21} f &= \partial_2 u \, \partial_1 f + \partial_1 u \, \partial_2 f + A_{21}, \\ \partial_{22} f &= -\partial_1 u \, \partial_1 f + \partial_2 u \, \partial_2 f + A_{22}. \end{aligned}$$

Observing that $e^{-u}\partial_1 f, e^{-u}\partial_2 f, \nu$ forms an orthonormal basis of $\mathbb{R}^3$, we get

$$|D^2 f|^2 = 4e^{2u}|Du|^2 + \sum_{i,j=1}^{2} |A_{ij}|^2.$$

Next $2\langle D\nu, \nu\rangle = D|\nu|^2 = 0$, hence $D\nu \in \operatorname{span}\{\partial_1 f, \partial_2 f\}$ and

$$\partial_i \nu = g^{kl}\langle \partial_i \nu, \partial_k f\rangle \partial_l f = -e^{-2u}\langle \nu, A_{i1}\rangle \partial_1 f - e^{-2u}\langle \nu, A_{i2}\rangle \partial_2 f.$$

This yields

$$|D\nu|^2 = e^{-2u} \sum_{i,j=1}^{2} |A_{ij}|^2$$

and

$$|D^2 f|^2 = 4e^{2u}|Du|^2 + e^{2u}|D\nu|^2.$$

□

4.C Huber's theorem for simply connected surfaces in codimension one

In this section we give a proof of Huber's theorem for simply connected surfaces in codimension one. Actually in Huber's theorem the assumption is only $\int_\Sigma K_- \, d\mu < \infty$ instead of $\int_\Sigma |A|^2 \, d\mu < \infty$, which is weaker by (1.1.2).

Theorem 4.C.1 ([Hu57]). *Let $f : \Sigma \to \mathbb{R}^3$ be a complete immersion of a simply connected surface Σ with square integrable second fundamental form $\int_\Sigma |A_f|^2 \, d\mu_f < \infty$. Then Σ with pull-back metric $f^* g_{euc}$ is conformally equivalent to the plane $\mathbb{C}$ or to the sphere S^2.*

Proof. By the uniformisation theorem for simply connected Riemann surfaces, see [FaKr, Theorem IV.4.1], Σ is conformally equivalent to the plane $\mathbb{C} = \mathbb{R}^2$, to the sphere S^2 or to the disc $B_1^2(0) \subseteq \mathbb{C}$. We have to exclude the disc.

Let's assume on contrary that there is a complete, conformal immersion $f : B_1(0) \to \mathbb{R}^3$ with pull-back metric $g = e^{2u} g_{euc} = f^* g_{euc}$ and with square integrable second fundamental form $\int_{B_1(0)} |A_f|^2 \, d\mu_f < \infty$. Obviously the metric $g = e^{2u} g_{euc}$ is complete on $B_1(0)$. Let $\nu : \mathbb{R}^2 \to S^2$ be the Gauß map determined up to the sign. Then as in the proof of Theorem 4.1.1 by elementary differential geometry

$$-\Delta u = K_g e^{2u} = *\nu^* \mathrm{vol}_{S^2} \quad \text{in } B_1(0). \tag{4.C.1}$$

We seek as in Proposition 4.1.1 a bounded solution v of $-\Delta v = *\nu^* \mathrm{vol}_{S^2}$ in a neigbourhood $\Omega := B_1(0) - \overline{B_\varrho(0)}$ of the boundary of $B_1(0)$. We get by (1.1.2) for ϱ close to 1 that

$$\begin{aligned} \mathrm{vol}_{S^2}(\varphi(\Omega)) &\le \int_\Omega J_{\mathbb{R}^2} \nu \, d\mathcal{L}^2 = \int_\Omega J\nu \, d\mu_g = \int_\Omega |K_g| \, d\mu_g \\ &\le \frac{1}{2} \int_\Omega |A_f|^2 \, d\mu_f < 4\pi \end{aligned}$$

in particular $\mathrm{vol}_{S^2}(S^2 - \varphi(\Omega)) > 0$. As $\varphi(\Omega)$ is a Borel set, actually σ-compact, in particular measurable, we can choose a compact subset $K \subseteq S^2 - \varphi(\Omega)$ with $\mathrm{vol}_{S^2}(K) > 0$ and put $U := S^2 - K \supseteq \varphi(\Omega)$ open.

By the construction as in Proposition 4.1.1, we get a smooth map $T : U \to B_2(0) \subseteq \mathbb{R}^2$ with

$$\begin{aligned} T^* dx_1 \wedge dx_2 &= \mathrm{vol}_S^2 | U, \\ \nabla^{S^2} T &\in L^\infty(U). \end{aligned}$$

Therefore $T \circ \nu \in W^{1,2}(\Omega, B_2(0))$ and extending by reflection we obtain $\Phi \in W^{1,2}(\mathbb{R}^2, \mathbb{R}^2)$ with

$$\nu^* \mathrm{vol}_{S^2} = (T \circ \nu)^* dx_1 \wedge dx_2 = \det D(T \circ \nu) dx_1 \wedge dx_2$$
$$= \det D\Phi \, dx_1 \wedge dx_2 \quad \text{in } \Omega.$$

By Theorem 4.A.2, we get $\det D\Phi \in \mathcal{H}^1(\mathbb{R}^2)$ and by Theorem 4.A.1 or by Theorem 4.A.3, there exists a continuous and bounded solution v of

$$-\Delta v = \det D\Phi \quad \text{in } \mathbb{R}^2,$$

in particular

$$-\Delta v = *\nu^* \mathrm{vol}_{S^2} = K_g e^{2u} \quad \text{in } \Omega.$$

By (4.C.1), we see that $u - v$ is harmonic in Ω. Putting

$$\alpha := \frac{1}{2\pi} \int_0^{2\pi} \langle \nabla(u - v)(r \cos\varphi, r \sin\varphi), (r \cos\varphi, r \sin\varphi) \rangle \, d\varphi \in \mathbb{R}$$
$$\text{for } \varrho < r < 1,$$

we see that $u - v - \alpha \log |.|$ is the real part of a holomorphic function Φ in Ω. By a Laurent expansion of Φ, we can write $\Phi = \Phi_0 + \Psi$ with Φ_0 holomorphic in $\mathbb{C} - \overline{B_\varrho(0)}$ and Ψ holomorphic in $B_1(0)$. We see that

$$\limsup_{|z| \nearrow 1} |u(z) - Re\Psi(z)| = \limsup_{|z| \nearrow 1} |v(z) + \alpha \log |z| + Re\Phi_0(z)| < \infty,$$

as v is bounded and Φ_0 is continuous around $\partial B_1(0)$. Recalling that the metric $e^{2u} g_{\mathrm{euc}}$ is complete on $B_1(0)$, we see that the metric $e^{2Re\Psi} g_{\mathrm{euc}}$ is complete on $B_1(0)$ too. Now $Re\Psi$ is harmonic on $B_1(0)$, and by the next Proposition a complete metric with harmonic conformal exponent does not exist on $B_1(0)$. This is a contradiction, and the theorem is proved. □

Proposition 4.C.1 ([Os69, Lemma 9.2]). *Let u be harmonic on a non-empty open subset $\Omega \subseteq \mathbb{C}$, such that the metric $e^{2u} g_{\mathrm{euc}}$ is complete on Ω. Then Ω is the plane $\mathbb{C}$ or the plane with one point deleted.*

Proof. By considering a connected component, we may assume that Ω is connected. As Ω is locally path-connected and locally simply connected, the universal covering $\pi : \Sigma \to \Omega$ exists and is simply connected by [Sp, Corollaries 2.5.7 and 2.5.14]. Σ with pull-back metric $\pi^* g_{\mathrm{euc}}$ is a simply connected Riemann surface, hence by the uniformisation theorem, see [FaKr, Theorem IV.4.1], it is conformally equivalent to the plane, sphere or disc. The sphere is excluded, since otherwise Ω were compact. Therefore we may assume $\Sigma = \mathbb{C}$ or $B_1(0)$ and have a conformal covering map $\pi : \Sigma \to \Omega$.

As $g := e^{2u} g_{\text{euc}}$ is complete on Ω and $\pi : \Sigma \to \Omega$ is a covering map, we claim that the pull-back metric $\pi^* g$ is complete on Σ. Indeed let $\gamma : [0, 1[\to \Sigma$ be a continuously differentiable path which leaves every compact subset of Σ. We have to prove that its length with respect to $\pi^* g$ is infinite. Now $\pi \circ \gamma$ is a continuously differentiable path in Ω with the same length as γ, as π is a local isometry. If $\pi \circ \gamma$ leaves any compact subset of Ω, it has infinite length by completness of g, and we are done. Otherwise there exists a sequence $t_n \nearrow 1$ with $(\pi \circ \gamma)(t_n) \to p \in \Omega$. We consider $B_\varrho(p) \subseteq \Omega$ which is evenly covered by π, that is the restriction of π to any component of $\pi^{-1}(B_\varrho(p))$ is a homeomorphsim onto $B_\varrho(p)$. If $\pi \circ \gamma$ eventually stays in $B_{\varrho/2}(p)$, that is $(\pi \circ \gamma)(t) \in B_{\varrho/2}(p)$ for $t \geq t_0$ for some $t_0 < 1$, we first obviously see that $\gamma(t) \in \pi^{-1}(B_\varrho(p))$ for $t \geq t_0$, hence $\gamma([t_0, 1[)$ being connected stays in one connected component C of $\pi^{-1}(B_\varrho(p))$. Then γ eventually stays in $C \cap \pi^{-1}(\overline{B_{\varrho/2}(p)})$, and as $\pi|C : C \xrightarrow{\approx} B_\varrho(p)$ is a homeomorphism, we see that $C \cap \pi^{-1}(\overline{B_{\varrho/2}(p)}) = (\pi|C)^{-1}(\overline{B_{\varrho/2}(p)})$ is compact, which contradicts the assumption that γ leaves any compact subset of Σ. Therefore $\pi \circ \gamma$ does not eventually stay in $B_{\varrho/2}(p)$. On the other hand $(\pi \circ \gamma)(t_n) \in B_{\varrho/4}(p)$ for n large. Combining these two facts, we get after relabeling the sequence two sequences $t_n < s_n < t_{n+1}$ with $(\pi \circ \gamma)(t_n) \in B_{\varrho/4}(p)$ and $(\pi \circ \gamma)(s_n) \notin B_{\varrho/2}(p)$. Then clearly $length((\pi \circ \gamma)|[t_n, s_n]) \geq c_0 > 0$, hence $\pi \circ \gamma$ has infinite length and the same for γ, proving the completness of $\pi^* g$.

As π is conformal and $d\pi$ has full rank, we can assume after a possible conjugation that $\pi : \Sigma \to \Omega \subseteq \mathbb{C}$ is holomorphic. Clearly,

$$\pi^* g = \pi^* e^{2u} g_{\text{euc}} = e^{2(u \circ \pi)} |\pi'|^2 g_{\text{euc}}.$$

Then as $\pi' \neq 0$, we see that $\log |\pi'|$ is harmonic on Σ. Moreover as u is harmonic, it is locally the real part of a holomorphic function φ, hence $u \circ \pi$ is locally the real part of the holomorphic function $\varphi \circ \pi$, and $u \circ \pi$ is harmonic on Σ. Then $v := (u \circ \pi) + \log |\pi'|$ is harmonic on Σ and $\pi^* g = e^{2v} g_{\text{euc}}$ is complete on Σ.

Since Σ is simply connected, $v = Re\Phi$ is the real part of a holomorphic function Φ globally on Σ. Again as Σ is simply connected and e^Φ is holomorphic on Σ, there exists a holomorphic function Ψ on Σ with $\Psi' = e^\Phi$. The pull-back metric of Ψ is given by

$$\Psi^* g_{\text{euc}} = |\Psi'|^2 g_{\text{euc}} = e^{2Re\Phi} g_{\text{euc}} = e^{2v} g_{\text{euc}} = \pi^* g$$

and hence is complete on Σ.

We endow Σ with the metric $\Psi^* g_{\text{euc}}$ and $\mathbb{C}$ with the Euclidean metric g_{euc}. Then obviously the map $\Psi : \Sigma \to \mathbb{C}$ is a local isometry, hence transforms geodesics on Σ into Euclidean geodesics on $\mathbb{C}$ that is into straight lines. We fix $p \in \Sigma$ and write to abbreviate the notation $\Psi(p) = 0 \in \mathbb{C}$. As Σ is complete the exponential map $\exp_p : T_p\Sigma \to \Sigma$ is defined on the whole of the tangent space and is surjective. We consider $\Psi \circ \exp_p : T_p\Sigma \to \mathbb{C}$ and see for any $v \in T_p\Sigma$ that $t \mapsto exp_p(tv)$ is a geodesic on Σ through p, hence by the remark above $t \mapsto (\Psi \circ \exp_p)(tv)$ is a stright line through 0 or likewise

$$(\Psi \circ \exp_p)(tv) = t \cdot \frac{d}{dt}(\Psi \circ \exp_p)(tv)|_{t=0} = t \cdot d(\Psi \circ \exp_p)_0.v,$$

hence $\Psi \circ \exp_p = d(\Psi \circ \exp_p)_0$. Recalling $d\exp_{p,0} = id_{T_p\Sigma}$, we obtain

$$\Psi \circ \exp_p = d(\Psi \circ \exp_p)_0 = d\Psi_p : T_p\Sigma \xrightarrow{\approx} \mathbb{C},$$

which is a linear isomorphism, as Ψ is a local isometry.

We conclude that $\exp_p$ is injective, and, as we already know that $\exp_p$ is surjective, we see that $\exp_p$ is bijective. Then so is $\Psi : \Sigma \xrightarrow{\approx} \mathbb{C}$, and hence the universal cover Σ of Ω is the plane, and we can consider the entire holomorphic function $\pi : \mathbb{C} \to \Omega \subseteq \mathbb{C}$. Now by the little Picard theorem, an entire holomorphic function excludes at most one point of $\mathbb{C}$, unless it is constant. As covering map π is surjective, in particular non-constant, hence $\Omega = \pi(\mathbb{C})$ is the plane or the plane with one point deleted, and the proposition is proved. □

References

[CLMSe93] R. Coifman, P. L. Lions, Y. Meyer and S. Semmes, *Compensated compactness and Hardy spaces*, Journal de Mathématiques pures et appliquées, **72** (1993), 247–286.

[FaKr] H. M. Farkas and I. Kra, "Riemann Surfaces", Springer Verlag, Berlin - Heidelberg - New York, 1991.

[FSt72] C. Fefferman and E. M. Stein, *H^p spaces of several variables*, Acta Mathematica, **129** (1972), 137–193.

[Hu57] A. Huber, *On subharmonic functions and differential geometry in the large*, Comment. Math. Helvetici, **32** (1957), 13–72.

[MuSv95] S. Müller and V. Šverák, *On surfaces of finite total curvature*, Journal of Differential Geometry, **42** (1995), 229–258.

[Os69] R. OSSERMAN, "A Survey of Minimal Surfaces", Van Nostrand, 1969.

[ScUh83] R. SCHOEN and K. UHLENBECK, *Boundary regularity and and the Dirichlet problem for harmonic maps*, Journal of Differential Geometry, **18** (1983), 253–268.

[Sp] E. H. SPANIER, "Algebraic Topology", McGraw Hill, 1966.

[We69] H. WENTE, *An existence theorem for surfaces of constant mean curvature*, Journal Math. Anal. Appl., **26** (1969), 318–344.

5 Removability of point singularities

5.1 $C^{1,\alpha}$-regularity for point singularities

We start with $C^{1,\alpha}$-regularity for point singularities with unit density of Willmore surfaces.

Lemma 5.1.1 ([KuSch04, Lemma 3.1]). *Let Σ be an open embedded Willmore surface in $\mathbb{R}^3$ with*

$$B_\delta(0) \cap \overline{\Sigma} - \Sigma = \{0\}, \tag{5.1.1}$$

$$\theta^{2,*}(\mathcal{H}^2 \llcorner \Sigma, 0) < 2, \tag{5.1.2}$$

$$\int_\Sigma |A_\Sigma|^2 \, d\mathcal{H}^2 < \infty. \tag{5.1.3}$$

Then $\Sigma \cup \{0\}$ is a $C^{1,\alpha}$-embedded surface at 0 for all $0 < \alpha < 1$, and the second fundamental form A satisfies the estimate

$$|A(x)| \le C_\varepsilon |x|^{-\varepsilon} \quad \forall \varepsilon > 0. \tag{5.1.4}$$

Proof. Clearly by (5.1.3)

$$\int_{\varrho^{-1}\Sigma \cap B_2(0)} |A_{\varrho^{-1}\Sigma}|^2 \, d\mathcal{H}^2 = \int_{\Sigma \cap B_{2\varrho}(0)} |A_\Sigma|^2 \, d\mathcal{H}^2 \to 0 \quad \text{for } \varrho \to 0.$$

Then for ϱ small, we get by interior estimates Theorem 3.3.1 that

$$\| \nabla^{\perp,k} A_{\varrho^{-1}\Sigma} \|_{L^\infty(B_1(0)-B_{1/2}(0))} \le C_k \| A_{\varrho^{-1}\Sigma} \|_{L^2(B_2(0)-B_{1/4}(0))} \to 0.$$

By (2.1.5) for $x \in B_1(0) - B_{1/2}(0), 0 < r \le 1/2$,

$$\begin{aligned} r^{-2}\mathcal{H}^2(B_r(x) \cap \varrho^{-1}\Sigma) &\le 8\mathcal{H}^2(B_{1/2}(x) \cap \varrho^{-1}\Sigma) \\ &\quad + C\mathcal{W}(B_{1/2}(x) \cap \varrho^{-1}\Sigma) \\ &\le 8\varrho^{-2}\mathcal{H}^2(B_{2\varrho}(0) \cap \Sigma) + C\mathcal{W}(B_{2\varrho}(x) \cap \Sigma) \end{aligned}$$

and

$$\limsup_{\varrho \to 0} r^{-2}\mathcal{H}^2(B_r(x) \cap \varrho^{-1}\Sigma) \le 8\pi\theta^{2,*}(\mathcal{H}^2 \llcorner \Sigma, 0) < \infty.$$

Similarly for other $x \in \mathbb{R}^3 - \{0\}$ and together $\varrho^{-1}\Sigma$ converges for subsequences smoothly on compact subsets of $\mathbb{R}^3 - \{0\}$ to a union of planes and

$$\mathcal{H}^2 \llcorner \varrho_j^{-1}\Sigma \to \mu := \sum_{i=1}^N \theta_i \mathcal{H}^2 \llcorner P_i$$

with $\theta_i \in \mathbb{N}$, $P_i \subseteq \mathbb{R}^3$ planes. As above

$$\begin{aligned} r^{-2}\mu(B_r(0)) &\leq \limsup_{\varrho\to 0} r^{-2}\mathcal{H}^2(B_r(0)\cap \varrho^{-1}\Sigma) \\ &= \limsup_{\varrho\to 0}(\varrho r)^{-2}\mathcal{H}^2(B_{\varrho r}(0)\cap\Sigma) \leq \pi\theta^{2,*}(\mathcal{H}^2 \llcorner \Sigma, 0) \end{aligned}$$

and

$$\sum_{i=1}^{N}\theta_i = \lim_{r\to\infty}\frac{\mu(B_r(0))}{\pi r^2} \leq \theta^{2,*}(\mathcal{H}^2\llcorner\Sigma, 0) < 2,$$

in particular $\mu = \mathcal{H}^2\llcorner P$. We claim $0 \in P = \operatorname{spt}\mu$. Indeed if not, then $B_{2r}(0)\cap \operatorname{spt}\mu = \emptyset$ for some $r > 0$, hence by smooth convergence $\varrho_j^{-1}\Sigma \cap \partial B_r(0) = \emptyset$ and likewise $\Sigma\cap\partial B_{r\varrho_j}(0) = \emptyset$ for j large. As $0 \in \overline{\Sigma}$, we see for $k \gg j$ that $\Sigma \cap B_{r\varrho_j}(0) - \overline{B_{r\varrho_k}(0)}$ is a non-empty, smooth, compact surface without boundary. By Proposition 2.1.1 (2.1.18)

$$4\pi \leq \mathcal{W}(\Sigma\cap B_{r\varrho_j}(0) - \overline{B_{r\varrho_k}(0)}) \leq 2\int_{\Sigma\cap B_{r\varrho_j}(0)} |A_\Sigma|^2 \, d\mathcal{H}^2,$$

which contradicts (5.1.3) for $j\to\infty$. Therefore $0 \in P = \operatorname{spt}\mu$, and $\varrho^{-1}\Sigma$ converges for subsequences to a unit-density plane through the origin, in particular $\theta^2(\mathcal{H}^2\llcorner\Sigma, 0) = 1$.

Further Σ is a smooth graph over some plane in $B_\varrho(0) - B_{\varrho/2}(0)$ for small ϱ, and hence it is a smooth embedded, unit-density Willmore surface in $B_\delta(0) - \{0\}$ for δ small enough which is diffeomorphic to an annulus

$$\Sigma \cap (B_\delta(0) - \{0\}) \cong B_1^2(0) - \{0\}.$$

We modify Σ outside $B_\delta(0)$ by a surface close to a plane such that Σ is a smooth, embedded surface in $\mathbb{R}^3 - \{0\}$ which is diffeomorphic to $\mathbb{R}^2 - \{0\}$, is Willmore in $B_\delta(0) - \{0\}$ and satisfies

$$\int_\Sigma |A_\Sigma|^2 \, d\mu_\Sigma < \varepsilon$$

for δ and ε small enough.

We invert Σ by $I(x) := |x|^{-2}x$ and put $\hat{\Sigma} := I(\Sigma)\cup\{0\}$. $\hat{\Sigma} \cong \mathbb{R}^2$ is a smooth, complete surface in $\mathbb{R}^3$. By invariance of the Willmore functional in (1.2.3), we get

$$\int_{\hat{\Sigma}} |A^0_{\hat{\Sigma}}|^2 \, d\mu_{\hat{\Sigma}} \leq \int_\Sigma |A_\Sigma|^2 \, d\mu_\Sigma < \varepsilon.$$

Next by Gauß-Bonnet's theorem and $\hat{\Sigma} \cong \mathbb{R}^2$

$$\lim_{R\to\infty} \int\limits_{\hat{\Sigma}\cap B_R(0)} K_{\hat{\Sigma}} \, d\mu_{\hat{\Sigma}} = 0,$$

hence by (1.1.7) and

$$\int\limits_{\hat{\Sigma}} |A_{\hat{\Sigma}}|^2 = 2 \int\limits_{\hat{\Sigma}} |A^0_{\hat{\Sigma}}|^2 + 2 \lim_{R\to\infty} \int\limits_{\hat{\Sigma}\cap B_R(0)} K_{\hat{\Sigma}} \, d\mu_{\hat{\Sigma}} = 2 \int\limits_{\hat{\Sigma}} |A^0_{\hat{\Sigma}}|^2 < 2\varepsilon,$$

in particular $A_{\hat{\Sigma}} \in L^2(\mu_{\hat{\Sigma}})$, hence $K_{\hat{\Sigma}} \in L^1(\mu_{\hat{\Sigma}})$ by (1.1.2) and

$$\int\limits_{\hat{\Sigma}} K_{\hat{\Sigma}} \, d\mu_{\hat{\Sigma}} = 0, \qquad \int\limits_{\hat{\Sigma}} |K_{\hat{\Sigma}}| \, d\mu_{\hat{\Sigma}} < \varepsilon.$$

Now $\hat{\Sigma}$ is a simply connected, complete, non-compact, oriented surface embedded in $\mathbb{R}^3$ with square integrable second fundamental form. By the uniformisation theorem for simply connected Riemann surfaces, see [FaKr] Theorem IV.4.1, it is conformally equivalent to the plane $\mathbb{C} = \mathbb{R}^2$ or to the disc $B^2_1(0) \subseteq \mathbb{C}$. Actually by Huber's theorem, see [Hu57] or Theorem 4.C.1, it is conformally equivalent to $\mathbb{C} = \mathbb{R}^2$, say

$$\hat{f} : \mathbb{R}^2 \xrightarrow{\approx} \hat{\Sigma} \subseteq \mathbb{R}^3$$

with pull-back metric $\hat{f}^* g_{\text{euc}} = e^{2\hat{u}} g_{\text{euc}}$. By Theorem 4.1.1 for $\varepsilon \leq 8\pi$, we see $\hat{u} \in L^\infty(\hat{\Sigma})$ and $\lim_{x\to\infty} |\hat{f}(x)|/|x| \in]0, \infty[$. Inverting back, we get a conformal diffeomorphism $f := I^{-1} \circ \hat{f} \circ I : (\mathbb{R}^2 \cup \{\infty\}) - \{0\} \xrightarrow{\cong} \Sigma$, as I is conformal, with pull-back metric $g = f^* g_{\text{euc}} = e^{2u} g_{\text{euc}}$ and

$$u \in L^\infty_{\text{loc}}(\mathbb{R}^2),$$

$$\lim_{y\to 0} \frac{|f(y)|}{|y|} \in]0, \infty[, \tag{5.1.5}$$

in particular there is $C < \infty$ such that

$$\Sigma \cap B_\varrho(0) \subseteq f(B^2_{C\varrho}(0)) \quad \text{for } \varrho > 0 \text{ small}.$$

f is a Willmore immersion near 0, say on $\Omega := B^2_1(0) - \{0\}$, hence it satisfies the Euler-Lagrange equation (1.3.10)

$$\Delta_g H + |A^0|^2 H = \delta\mathcal{W}(f) \nu_\Sigma = 0 \quad \text{in } \Omega,$$

where ν_Σ denotes a unit normal at Σ, and by conformal transformation of the laplacian

$$\Delta H + e^{2u}|A^0|^2 H = 0 \quad \text{in } \Omega.$$

We want to apply the Power-decay-Lemma 5.2.1 of the next section to $v = H$. Clearly

$$|v|, e^u|A^0| \leq C|A| \quad \text{in } \Omega,$$

and

$$A \in L^2(B_1^2(0)).$$

This verifies (5.2.1), (5.2.2) and (5.2.4). By interior estimates Theorem 3.3.1, by (5.1.5) above and since f is Willmore in Ω, we get for $\int_\Omega |A|^2 \, \mathrm{d}\mu_g < \varepsilon_1$ that

$$\| A \|_{L^\infty(B_\varrho^2)} \leq C\varrho^{-1} \| A \|_{L^2(B_{2\varrho}^2)} \quad \text{for any } B_{2\varrho}^2 \subseteq \Omega. \tag{5.1.6}$$

This verifies (5.2.3), and the Power-decay-Lemma 5.2.1 implies

$$\int\limits_{B_\varrho(0)\cap\Sigma} |\vec{\mathbf{H}}_\Sigma|^2 \, \mathrm{d}\mathcal{H}^2 \leq \int\limits_{B_{C\varrho}^2(0)} |H|^2 \, \mathrm{d}\mu_g \leq C_\varepsilon \varrho^{2-\varepsilon} \quad \forall \varepsilon > 0. \tag{5.1.7}$$

We approximate f by affine functions $l(y) := a + by$ and put

$$\omega(\varrho) := \inf_{l \text{ affine}} \varrho^{-2} \| f - l \|_{L^\infty(B_\varrho^2(0))}^2 \quad \text{for } 0 < \varrho \leq 1.$$

Putting $f_\varrho(y) := \varrho^{-1} f(\varrho y)$, we see by rescaling that

$$\omega(\varrho) = \inf_{l \text{ affine}} \| f_\varrho - l \|_{L^\infty(B_1^2(0))}^2 .$$

Now we use

$$\Delta f = e^{2u}\vec{\mathbf{H}} =: h \quad \text{on } \Omega, \tag{5.1.8}$$

or likewise putting $h_\varrho(y) := \varrho h(\varrho y)$ that $\Delta f_\varrho = h_\varrho$. By standard elliptic theory, there exists a solution $\phi_\varrho \in W^{2,2}(B_1^2(0))$

$$\Delta\phi_\varrho = h_\varrho \quad \text{in } B_1^2(0), \quad \phi_\varrho = 0 \quad \text{on } \partial B_1^2(0),$$

and this solution satisfies by Friedrich's theorem, see [GT,Theorem 8.12], and (5.1.7) that

$$\begin{aligned} \| \phi_\varrho \|_{L^\infty(B_1^2(0))} &\leq C \| \phi_\varrho \|_{W^{2,2}(B_1^2(0))} \leq C \| h_\varrho \|_{L^2(B_1^2(0))} \\ &= C \| h \|_{L^2(B_\varrho^2(0))} \leq C_\varepsilon \varrho^{1-\varepsilon} \quad \forall \varepsilon > 0. \end{aligned}$$

Clearly $f_\varrho - \phi_\varrho$ is harmonic in Ω and bounded in $B^2_1(0)$, hence $f_\varrho - \phi_\varrho$ is harmonic in $B^2_1(0)$ and so is $f_\varrho - l_\varrho - \phi_\varrho$ for any affine function l_ϱ. We see for the linear Taylor polynom $l(y) := (f_\varrho - l_\varrho - \phi_\varrho)(0) + \nabla(f_\varrho - l_\varrho - \phi_\varrho)(0)y$ and $0 < \sigma < 1/2$ by Cauchy-estimates

$$\begin{aligned}\sigma^{-2} \parallel f_\varrho - l_\varrho - \phi_\varrho - l \parallel_{L^\infty(B^2_\sigma(0))} &\leq \parallel D^2(f_\varrho - l_\varrho - \phi_\varrho) \parallel_{L^\infty(B^2_{1/2}(0))} \\ &\leq C \parallel f_\varrho - l_\varrho - \phi_\varrho \parallel_{L^\infty(B^2_1(0))} \\ &\leq C \parallel f_\varrho - l_\varrho \parallel_{L^\infty(B^2_1(0))} + C_\varepsilon \varrho^{1-\varepsilon} \ \forall \varepsilon > 0.\end{aligned}$$

By rescaling as above

$$\begin{aligned}\omega(\sigma\varrho) &\leq \sigma^{-2} \parallel f_\varrho - l_\varrho - l \parallel^2_{L^\infty(B^2_\sigma(0))} \\ &\leq 2\sigma^{-2} \parallel f_\varrho - l_\varrho - \phi_\varrho - l \parallel^2_{L^\infty(B^2_\sigma(0))} + 2\sigma^{-2} \parallel \phi_\varrho \parallel^2_{L^\infty(B^2_1(0))} \\ &\leq C\sigma^2 \parallel f_\varrho - l_\varrho \parallel^2_{L^\infty(B^2_1(0))} + C_\varepsilon \sigma^{-2} \varrho^{2-\varepsilon}\end{aligned}$$

and taking the infimum over l_ϱ, we arrive at

$$\omega(\sigma\varrho) \leq C\sigma^2 \omega(\varrho) + C_\varepsilon \sigma^{-2} \varrho^{2-\varepsilon}.$$

Then standard iteration yields

$$\omega(\varrho) \leq C_\varepsilon \varrho^{2-\varepsilon} \quad \forall \varepsilon > 0.$$

We have observed above that $f - \phi_1$ is harmonic, in particular smooth, hence $f \in W^{2,2}(B^2_1(0))$ and $\Delta f = e^{2u}\vec{\mathbf{H}}$ almost everywhere in $B^2_1(0)$ by (5.1.8). Then by Friedrich's theorem, see [GT, Theorem 8.8], and (5.1.7) for appropriate affine $l_{2\varrho}$

$$\int_{B^2_\varrho(0)} |D^2 f|^2 \, d\mathcal{L}^2 \leq C \int_{B^2_{2\varrho}(y)} |\Delta f|^2 \, d\mathcal{L}^2 + C\varrho^{-4} \int_{B^2_{2\varrho}(y)} |f - l_{2\varrho}|^2 \, d\mathcal{L}^2 \leq C_\varepsilon \varrho^{2-\varepsilon}.$$

Recalling $A_{ij} = (\partial_{ij} f)^\perp$ is the normal projection, we get

$$\parallel A \parallel_{L^2(B_\varrho(0))} \leq C_\varepsilon \varrho^{1-\varepsilon},$$

and by interior estimates Theorem 3.3.1

$$\parallel A \parallel_{L^\infty(\partial B_\varrho(0))} \leq C\varrho^{-1} \parallel A \parallel_{L^2(B_{2\varrho}(0))} \leq C_\varepsilon \varrho^{-\varepsilon},$$

which is (5.1.4).

(5.1.4), (5.1.5) and (5.1.8) imply $|\Delta f(y)| \leq C_\varepsilon |y|^{-\varepsilon}$ for all $\varepsilon > 0$, hence $\Delta f \in L^p(B^2_1(0))$ for all $1 \leq p < \infty$ and $f \in W^{2,p}(B^2_1(0)) \hookrightarrow$

$C^{1,1-\varepsilon}(B_1^2(0))$. As the pull-back metric $f^* g_{\text{euc}} = e^{2u} g_{\text{euc}}, u \in L^\infty$, does not degenerate in 0, we see that $Df(0)$ has full rank, and $f : B_1^2(0) \to \Sigma \cup \{0\}$ is a $C^{1,1-\varepsilon}$-immersion and $\Sigma \cup \{0\}$ is a $C^{1,1-\varepsilon}$-embedded surface at 0 for all $\varepsilon > 0$, concluding the proof. □

Remarks 5.1.

1. The above lemma cannot be improved to get $C^{1,1}$-regularity. Indeed, the inverted catenoid is a Willmore surface as it is an inversion of a minimal surface. Like the catenoid, it has square integrable second fundamental form. It admits the parametrisation

$$f(t,\theta) = \frac{\cosh t}{\cosh(t)^2 + t^2}(\cos\theta, \sin\theta, 0) \pm \frac{t}{\cosh(t)^2 + t^2}\, e_3$$

and consists of two graphs near 0 which correspond to $\pm t > 0$. Therefore each of these graphs satisfy the assumptions of lemma near 0. Writing $r = \sqrt{x^2 + y^2} = \cosh t / (\cosh(t)^2 + t^2)$, we see

$$\varphi(r) = \frac{\pm t}{\cosh(t)^2 + t^2} \approx \pm r^2 \log \frac{1}{r},$$

hence these graphs are not $C^{1,1}$ near 0.
2. Riviere has extended the above lemma to any codimension, see [Ri08].

5.2 Power-decay

Lemma 5.2.1 (Power-decay-Lemma [KuSch04, Lemma 2.1]). *Let $v \in C^\infty(\Omega)$, $\Omega := B_1^2(0) - \{0\} \subseteq \mathbb{R}^2$, A be measurable on Ω satisfying*

$$|\Delta v| \leq |A|^2\, |v| \quad \text{in } \Omega, \tag{5.2.1}$$

$$|v| \leq C|A| \quad \text{in } \Omega, \tag{5.2.2}$$

$$\| A \|_{L^\infty(B_\varrho)} \leq C\varrho^{-1} \| A \|_{L^2(B_{2\varrho})} \quad \text{for} \quad B_{2\varrho} \subseteq \Omega, \tag{5.2.3}$$

$$\int_\Omega |A|^2 < \infty. \tag{5.2.4}$$

Then $\forall\, \varepsilon > 0 : \exists\, C_\varepsilon < \infty : \forall\, 0 < \varrho \leq 1 :$

$$\int_{B_\varrho(0)} |v|^2 \leq C_\varepsilon \varrho^{2-\varepsilon}. \tag{5.2.5}$$

Remark 5.1. From (5.2.1) - (5.2.4), we can conclude

$$\begin{aligned}\Delta v + qv = 0 \quad \text{in } B_1^2(0) - \{0\},\\ |y|^2 q(y) \to 0 \text{ for } y \to 0.\end{aligned} \tag{5.2.6}$$

In [Sim96] equations with this asymptotics were investigated, and Lemma 1.4 in [Sim96] yields

$$\begin{aligned}\varrho^{-1} \parallel v \parallel_{L^2(B_\varrho(0)-B_{\varrho/2}(0))} &= O(\varrho^{k+\varepsilon}) \Longrightarrow \varrho^{-1} \parallel v \parallel_{L^2(B_\varrho(0)-B_{\varrho/2}(0))}\\ &= O(\varrho^{k+1-\varepsilon})\end{aligned}$$

for all $k \in \mathbb{Z}, \varepsilon > 0$. From (5.2.2) we only get $v(y) = o(|y|^{-1})$ which does not suffice to obtain the conclusion (5.2.5) from (5.2.6) as the example

$$v(y) = v(r(\cos\varphi, \sin\varphi)) := \frac{1}{r\log(2/r)}\cos\varphi$$

shows. For the proof of the Power-decay-Lemma it is decisive to observe that

$$\int_0^{1/2} \sup_{|y|=\varrho} |q(y)|\varrho \, d\varrho < \infty$$

by (5.2.3) and (5.2.4).

5.3 Higher regularity for point singularities

Let Σ be an open surface and $f_t : \Sigma \to \mathbb{R}^n$ be a smooth family of immersions with

$$\partial_t f_t|_{t=0} = V =: N + Df.\xi$$

where $N \in N\Sigma$ is normal and $\xi \in T\Sigma$ is tangential. The first variation of the Willmore integrand was calculated for normal variations $V = N$ in (1.3.5) to be

$$\begin{aligned}\partial_t\Big(\frac{1}{4}|\vec{\mathbf{H}}_{f_t}|^2\mu_{f_t}\Big) &= \frac{1}{2}\langle\Delta^\perp V + Q(A^0)V, \vec{\mathbf{H}}\rangle\mu_f\\ &= \frac{1}{2}\langle\Delta^\perp\vec{\mathbf{H}} + Q(A^0)\vec{\mathbf{H}}, N\rangle\mu_f\\ &\quad + \frac{1}{2}g^{ij}\nabla_i\Big(\langle\nabla_j^\perp N, \vec{\mathbf{H}}\rangle - \langle N, \nabla_j^\perp\vec{\mathbf{H}}\rangle\Big)\mu_f.\end{aligned} \tag{5.3.1}$$

For tangential variations $V = Df.\xi$, we consider the flow ϕ_t of ξ, that is $\phi_0 = id_\Sigma$, $\partial_t \phi_t = \xi \circ \phi_t$ and calculate for $t = 0$

$$\begin{aligned}
\partial_t\Big(\frac{1}{4}|\vec{\mathbf{H}}_{f_t}|^2\mu_{f_t}\Big) &= \partial_t\Big(\frac{1}{4}|\vec{\mathbf{H}}_{f\circ\phi_t}|^2\mu_{f\circ\phi_t}\Big) \\
&= \partial_t\Big(\frac{1}{4}|\vec{\mathbf{H}}\circ\phi_t|^2\ \phi_t^*\mu_f\Big) \\
&= \partial_t\Big(\frac{1}{4}|\vec{\mathbf{H}}\circ\phi_t|^2\ \sqrt{\phi_t^* g}\mathcal{L}^2\Big) \\
&= \frac{1}{4}\xi^i\partial_i|\vec{\mathbf{H}}|^2\mu_f + \frac{1}{4}|\vec{\mathbf{H}}|^2\ \partial_t\sqrt{\phi_t^* g}\ \mathcal{L}^2.
\end{aligned}$$

Here $\phi_t^* g_{ij} = (g_{kl}\circ\phi_t)\partial_i\phi_t^k\partial_j\phi_t^l$ and putting $g_{t,ij} = g_{kl}\partial_i\phi_t^k\partial_j\phi_t^l$, we calculate

$$\begin{aligned}
\partial_t\sqrt{\phi_t^* g} &= \partial_t\sqrt{g\circ\phi_t} + \partial_t g_t/(2\sqrt{g_t}) = \xi^i\partial_i\sqrt{g} + \frac{1}{2}\partial_t g_{t,ij}g^{ij}\sqrt{g} \\
&= \xi^i\partial_i\sqrt{g} + g_{kl}\partial_i\xi^k\delta_j^l g^{ij}\sqrt{g} = \xi^i\partial_i\sqrt{g} + \partial_i\xi^i\sqrt{g} = \partial_i(\xi^i\sqrt{g}).
\end{aligned}$$

Together we get

$$\partial_t\Big(\frac{1}{4}|\vec{\mathbf{H}}_{f_t}|^2\mu_{f_t}\Big) = \partial_i\Big(\frac{1}{4}|\vec{\mathbf{H}}|^2\xi^i\sqrt{g}\Big)\mathcal{L}^2.$$

Observing

$$\delta_k^i\Gamma_{ij}^k = \frac{1}{2}g^{il}(\partial_i g_{jl} + \partial_j g_{li} - \partial_l g_{ij}) = \frac{1}{2}g^{il}\partial_j g_{li} = \frac{1}{2}\partial_j g/g = \partial_j\sqrt{g}/\sqrt{g},$$

we calculate for any tangential η that

$$\nabla_i\eta^i = \delta_k^i\nabla_i\eta^k = \partial_i\eta^i + \delta_k^i\Gamma_{ij}^k\eta^j = \partial_i\eta_i + \eta^i\partial_i\sqrt{g}/\sqrt{g} = \frac{1}{\sqrt{g}}\partial_i(\eta^i\sqrt{g})$$

and get

$$\partial_t\Big(\frac{1}{4}|\vec{\mathbf{H}}_{f_t}|^2\mu_{f_t}\Big) = \nabla_i\Big(\frac{1}{4}|\vec{\mathbf{H}}|^2\xi^i\Big)\mu_f. \tag{5.3.2}$$

Moreover for η with compact support in Σ

$$\int_\Sigma \nabla_i\eta^i \mathrm{vol}_g = \int_\Sigma \partial_i(\eta^i\sqrt{g})d\mathcal{L}^2 = 0. \tag{5.3.3}$$

Abbreviating

$$\sigma_j := \frac{1}{2}\langle\nabla_j^\perp N, \vec{\mathbf{H}}\rangle - \frac{1}{2}\langle N, \nabla_j^\perp\vec{\mathbf{H}}\rangle + \frac{1}{4}|\vec{\mathbf{H}}|^2 g_{jk}\xi^k, \tag{5.3.4}$$

we get combining (5.3.1) and (5.3.2)

$$\partial_t\Big(\frac{1}{4}|\vec{\mathbf{H}}_{f_t}|^2\mu_{f_t}\Big) = \frac{1}{2}\langle \Delta^\perp\vec{\mathbf{H}} + Q(A^0)\vec{\mathbf{H}}, N\rangle\mu_f + g^{ij}\nabla_i\sigma_j\,\mu_f. \quad (5.3.5)$$

Now we consider the 1-form $\sigma_V := \sigma_j dx^j$ on Σ and its hodge $\omega_V := *\sigma_V$ with respect to g. By Stokes theorem and (5.3.3), we calculate for any $\psi \in C_0^1(\Sigma)$

$$\begin{aligned}\int_\Sigma \psi\,\mathrm{d}\omega_V &= -\int_\Sigma \mathrm{d}\psi \wedge *\sigma_V = -\int_\Sigma \langle\,\mathrm{d}\psi, \sigma_V\rangle \mathrm{vol}_g \\ &= -\int_\Sigma g^{ij}\partial_i\psi\,\sigma_j\,\mathrm{vol}_g = \int_\Sigma \psi g^{ij}\nabla_i\sigma_j \mathrm{vol}_g,\end{aligned}$$

hence

$$\mathrm{d}\omega_V = g^{ij}\nabla_i\sigma_j\,\mathrm{vol}_g$$

and by (5.3.5)

$$\partial_t\Big(\frac{1}{4}|H|^2\,\mathrm{d}\mu\Big) = \frac{1}{2}\langle \Delta_g\vec{\mathbf{H}} + Q(A^0)\vec{\mathbf{H}}, V\rangle\,\mathrm{d}\mu + d\omega_V. \quad (5.3.6)$$

Proposition 5.3.1. *If $f : \Sigma \to \mathbb{R}^n$ is Willmore and $V \equiv \mathrm{const} \in \mathbb{R}^n$ then ω_V is a closed 1-form on Σ.*

Proof. We obtain for any open $\Omega \subseteq \Sigma$

$$0 = \frac{\mathrm{d}}{\mathrm{d}t}\int_\Omega \partial_t\Big(\frac{1}{4}|\vec{\mathbf{H}}_{f+tV}|^2\mu_{f+tV}\Big) = \int_\Omega \mathrm{d}\omega_V,$$

hence ω_V is closed on Σ. ☐

Lemma 5.3.1. *Let $\Sigma \subseteq \mathbb{R}^n$ be an open embedded $C^{1,\alpha}$-surface, $0 < \alpha < 1$, with $0 \in \Sigma$, $\Sigma - \{0\}$ a smooth Willmore surface and*

$$|A(x)| \le C_\varepsilon |x|^{-\varepsilon} \quad \forall \varepsilon > 0. \quad (5.3.7)$$

Then we have the expansion

$$\vec{\mathbf{H}}(x) = \vec{\mathbf{H}}_0 \log|x| + C^{0,\alpha}_{\mathrm{loc}}, \quad \nabla^\perp\vec{\mathbf{H}}(x) = \frac{\vec{\mathbf{H}}_0 x^T}{|x|^2} + O(|x|^{\alpha-1}) \quad (5.3.8)$$

for some $\vec{\mathbf{H}}_0 \in N_0\Sigma \subseteq \mathbb{R}^n$ which we call the residue

$$\mathrm{Res}_\Sigma(0) := \vec{\mathbf{H}}_0$$

of Σ at 0.

The residue can be calculated with the use of the closed 1-form ω_V *on* $\Sigma - \{0\}$ *for any* $V \in \mathbb{R}^n$ *by*

$$\int_{\partial \Sigma_\varrho} \omega_V \to -\pi \langle V, \mathrm{Res}_\Sigma(0) \rangle \quad \text{for } \varrho \to 0, \tag{5.3.9}$$

where $\Sigma_\varrho := B_\varrho(0) \cap \Sigma$.

If $\mathrm{Res}_\Sigma(0) = 0$ *then* Σ *is a smooth Willmore surface.*

Proof. Since the induced metric is $C^{0,\alpha}$, we can choose by standard elliptic theory a local conformal $C^{1,\alpha}$-parametrization $f : B_1^2(0) \xrightarrow{\approx} \Sigma$ around $0 = f(0)$ with conformal factor $|\partial_i f|^2 =: e^{2u}$ and $Df(0) = i : \mathbb{R}^2 \hookrightarrow \mathbb{R}^n$.

Since $\Sigma - \{0\}$ is Willmore, we get the Euler-Lagrange equation (1.3.10) which reads in local conformal coordinates

$$\Delta \vec{\mathbf{H}} = -e^{2u} Q(A^0) \vec{\mathbf{H}} \in L^p(B_1^2(0)) \quad \forall p < \infty.$$

By standard elliptic theory there exists a solution $\phi \in W^{2,p}(B_1^2(0)) \hookrightarrow C^{1,\alpha}(B_1^2(0))$

$$\Delta \phi = -e^{2u} Q(A^0) \vec{\mathbf{H}} \quad \text{in } B_1^2(0), \qquad \phi = 0 \quad \text{on } \partial B_1^2(0). \tag{5.3.10}$$

We see that $\vec{\mathbf{H}} - \phi$ is harmonic in $B_1^2(0) - \{0\}$, and, as $|\vec{\mathbf{H}}(y) - \phi(y)| \leq C_\varepsilon |y|^{-\varepsilon}$, the only singular contribution can be a logarithm, hence

$$\vec{\mathbf{H}}(y) = \vec{\mathbf{H}}_0 \log|y| + C^{1,\alpha}_{\mathrm{loc}}$$

for some $\vec{\mathbf{H}}_0 \in \mathbb{R}^n$. Since $\vec{\mathbf{H}}(y) \in N\Sigma_{f(y)}$, we see

$$0 = \lim_{y \to 0} \langle \partial_i f(y), \vec{\mathbf{H}}(y) \rangle / \log|y| = \langle \partial_i f(0), \vec{\mathbf{H}}_0 \rangle,$$

hence $\vec{\mathbf{H}}_0 \in N\Sigma_0$. Differentiating, we get

$$\nabla \vec{\mathbf{H}}(y) = \frac{H_0 y^T}{|y|^2} + C^{0,\alpha}_{\mathrm{loc}}.$$

Recalling that $f \in C^{1,\alpha}$ and $Df(0) = i : \mathbb{R}^2 \hookrightarrow \mathbb{R}^n$, hence $x = f(y) = y + O(|y|^{1+\alpha})$, we arrive at (5.3.8).

When the residue $\vec{\mathbf{H}}_0$ vanishes, we see that $\vec{\mathbf{H}} - \phi$ is harmonic in $B_1^2(0)$, hence $\vec{\mathbf{H}} \in C^{1,\alpha}_{\mathrm{loc}}(B_1^2(0))$. In general, we see from the equation

$$\Delta f = e^{2u} \vec{\mathbf{H}} \quad \text{weakly in } B_1^2(0)$$

and $f \in C^{1,\alpha}(B_1^2(0))$, $e^{2u} = |\partial_i f|^2$ that $\vec{\mathbf{H}} \in C^{k,\alpha}_{\text{loc}}, k \geq 0$ implies $f \in C^{k+2,\alpha}_{\text{loc}}$ and $A = \nabla^2 f \in C^{k,\alpha}_{\text{loc}}$. This in turn yields $\phi \in C^{k+2,\alpha}_{\text{loc}}$ and $\vec{\mathbf{H}} \in C^{k+2,\alpha}_{\text{loc}}$. Then the bootstrap proceeds proving that f and Σ are smooth.

Finally, we calculate the residue with the help of ω_V. For $0 < \varrho \ll 1$ small, we see that $\Sigma_\varrho := B_\varrho(0) \cap \Sigma$ is a disc whose boundary $\partial \Sigma_\varrho = \partial B_\varrho(0) \cap \Sigma$ is a smooth curve converging when rescaled to a planar circle as $\Sigma \in C^{1,\alpha}$. More precisely, we get for the unit outward normal at $\partial \Sigma_\varrho$ in Σ

$$n_\varrho(x) = \frac{x}{|x|} + O(|x|^\alpha). \tag{5.3.11}$$

As $*n_\varrho$ is the positive oriented tangent of $\partial \Sigma_\varrho$, we see

$$\int_{\partial \Sigma_\varrho} \omega_V = \int_{\partial \Sigma_\varrho} \omega_V(*n_\varrho)\, d\mathcal{H}^1 = -\int_{\partial \Sigma_\varrho} (*\omega_V)(n_\varrho)\, d\mathcal{H}^1 = \int_{\partial \Sigma_\varrho} \sigma_V(n_\varrho)\, d\mathcal{H}^1.$$

Decomposing $V =: N + \xi$, $N \in N\Sigma$, $\xi \in T\Sigma$, in normal and tangential components, we calculate by the definition (5.3.4)

$$\sigma_V(n_\varrho) = \sigma_j n_\varrho^j = \frac{1}{2}\langle \nabla^\perp_{n_\varrho} N, \vec{\mathbf{H}}\rangle - \frac{1}{2}\langle N, \nabla^\perp_{n_\varrho} \vec{\mathbf{H}}\rangle + \frac{1}{4}|\vec{\mathbf{H}}|^2 g_{jk} n_\varrho^j \xi^k.$$

Using (5.3.7), we estimate

$$\left| \int_{\partial \Sigma_\varrho} \frac{1}{4}|\vec{\mathbf{H}}|^2 g_{jk} n_\varrho^j \xi^k \, d\mathcal{H}^1 \right| \leq C\varrho C_\varepsilon \varrho^{-\varepsilon} \to 0,$$

and

$$|\langle \nabla^\perp_{n_\varrho} N, \vec{\mathbf{H}}\rangle| = |\langle \nabla^\perp_{n_\varrho} (V - \xi), \vec{\mathbf{H}}\rangle| = |\langle A(\xi, n_\varrho), \vec{\mathbf{H}}\rangle| \leq C_\varepsilon \varrho^{-\varepsilon},$$

hence

$$\left| \int_{\partial \Sigma_\varrho} \frac{1}{2}\langle \nabla^\perp_{n_\varrho} N, \vec{\mathbf{H}}\rangle \, d\mathcal{H}^1 \right| \to 0.$$

From (5.3.7), (5.3.8) and (5.3.11), we obtain

$$\begin{aligned} \langle N, \nabla^\perp_{n_\varrho} \vec{\mathbf{H}}\rangle &= \left\langle N, \left(\vec{\mathbf{H}}_0 \frac{x^T}{|x|^2} + O(|x|^{\alpha-1})\right)\left(\frac{x}{|x|} + O(|x|^\alpha)\right)\right\rangle \Big| \\ &= \langle N, \vec{\mathbf{H}}_0\rangle \frac{1}{|x|} + O(|x|^{\alpha-1}), \end{aligned}$$

hence

$$\int_{\partial \Sigma_\varrho} \frac{1}{2} \langle N, \nabla^\perp_{n_\varrho} \vec{\mathbf{H}} \rangle \, d\mathcal{H}^1 \to \pi \langle V, \vec{\mathbf{H}}_0 \rangle,$$

and (5.3.9) follows. □

Lemma 5.3.2. *Let* Σ *be an open embedded Willmore surface in* $\mathbb{R}^3$ *with*

$$\overline{\Sigma} - \Sigma = \{p_1, \ldots, p_N\}$$

$$\theta^{2,*}(\mathcal{H}^2 \llcorner \Sigma, p_k) < 2, \tag{5.3.12}$$

$$\int_\Sigma |A_\Sigma|^2 \, d\mathcal{H}^2 < \infty. \tag{5.3.13}$$

Then $\overline{\Sigma} = \Sigma \cup \{p_1, \ldots, p_k\}$ *is* $C^{1,\alpha}$*-embedded with unit-density near* p_k *and*

$$\sum_{k=1}^N \mathrm{Res}_{\overline{\Sigma}}(p_k) = 0, \tag{5.3.14}$$

where the Residue $\mathrm{Res}_{\overline{\Sigma}}$ *is defined in Lemma* 5.3.1.

In particular, if $N = 1$ *then* $\overline{\Sigma}$ *is a smooth, embedded Willmore surface.*

Proof. By Lemma 5.1.1 $\overline{\Sigma}$ is a $C^{1,\alpha}$-embedded, unit-density surface satisfying (5.3.7) near p_k, and the residue of Σ at p_k is well defined. Putting

$$\Omega_\varrho := \Sigma - \cup_{k=1}^N B_\varrho(p_k) \Subset \Sigma$$

and $\Sigma_\varrho(p_k) := B_\varrho(p_k) \cap \Sigma$ for small $\varrho > 0$, we obtain for any $V \in \mathbb{R}^n$ and the associated closed 1-form ω_V on Σ in (5.3.4) that

$$0 = \int_{\Omega_\varrho} d\omega_V = -\sum_{k=1}^N \int_{\partial \Sigma_\varrho(p_k)} \omega_V \to \pi \langle V, \sum_{k=1}^N \mathrm{Res}_{\overline{\Sigma}}(p_k) \rangle,$$

hence

$$\sum_{k=1}^N \mathrm{Res}_{\overline{\Sigma}}(p_k) = 0,$$

as V is arbitrary.

When $N = 1$, this means $\mathrm{Res}_{\overline{\Sigma}}(p_1) = 0$, and $\overline{\Sigma}$ is a smooth, embedded Willmore surface according to Lemma 5.3.1. □

Remark 5.2. By Riviere's result in [Ri08], the above lemma holds in any codimension.

5.4 Applications

In the following applications, we will strongly use Bryant's result in [Bry82] that Willmore spheres $M^2 \subseteq \mathbb{R}^3$ which are not round spheres satisfy

$$\mathcal{W}(M^2) \geq 16\pi. \tag{5.4.1}$$

A more elementary proof of [Bry82] Theorem E can be found in [Es88] §6 Proposition. When combined with a theorem of Osserman [Os69] Theorem 9.2, one obtains the estimate slightly weaker than (5.4.1) that Willmore spheres $M^2 \subseteq \mathbb{R}^3$ which are not round spheres satisfy

$$\mathcal{W}(M^2) \geq 8\pi.$$

Actually, this estimate suffices for all applications in this section, except that we have to assume the strict inequality

$$\mathcal{W}(f_0) < 8\pi$$

in (5.4.3) below.

Proposition 5.4.1. *Let* $f_t : \Sigma \to \mathbb{R}^3$ *be a Willmore flow of a closed surface* Σ *with*

$$\mathcal{W}(f_0) \leq 8\pi.$$

Then non-compact limits under the blow up procedure in Section 3.3 are connected, smooth, after inversion, and are never spheres, in case they are non-trivial.

Proof. If f_0 is Willmore then the flow is constant and all limits are the initial surface itself up to translation, which are all compact. If f_0 is not Willmore then the flow is not stationary, hence $\mathcal{W}(f_t) < 8\pi$ for $t > 0$ and when considering the flow $f_{\tau+t}$ for some $\tau > 0$, we may assume that

$$\mathcal{W}(f_t) \leq 8\pi - \delta \text{ for some } \delta > 0.$$

Let $f^\infty : \Sigma^\infty \to \mathbb{R}^3$ be a limit under the blow up procedure of Section 3.3, which is a smooth, proper Willmore immersion, for example

$$f^j := \varrho_{t_j}^{-1}(f_{t_j + c_0 \varrho_{t_j}^4} - y_{t_j}) \to f^\infty$$

smoothly after reparametrization on compact subsets of $\mathbb{R}^3$. Clearly $f^j(\mu_{f^j}) \to f^\infty(\mu_{f^\infty}) := \mu_\infty$ weakly as Radon measures and by (2.1.24) that

$$\begin{aligned}\#(f^{\infty,-1}(.)) &\leq \theta^2(\mu_\infty, \cdot) \leq \theta^2(\mu_\infty, \infty) + \frac{1}{4\pi}\mathcal{W}(f^\infty) \\ &\leq \liminf_{j\to\infty} \frac{1}{4\pi}\mathcal{W}(f^j) \leq 2 - \delta/(4\pi) < 2.\end{aligned}$$

We conclude that f^∞ is an embedding, that $\Sigma^\infty := f^\infty(\Sigma^\infty) \subseteq \mathbb{R}^3$ is a smooth embedded Willmore surface, and that Σ^∞ is connected by Proposition 2.1.4 (2.1.22).

We select $x_0 \notin \Sigma^\infty$, hence $\overline{B_{R^{-1}(x_0)}} \cap \Sigma^\infty = \emptyset$ for some $R < \infty$ and consider the inversion $I(x) := |x - x_0|^{-2}(x - x_0)$. Then $\hat{\Sigma}^\infty := I(\Sigma^\infty) \subseteq B_R(0) - \{0\}$ is a smooth Willmore surface. We address the question of smoothness of Σ^∞ at infinity, or more precisely whether $\hat{\Sigma}^\infty$ is smooth at 0. By above

$$\theta^2(\mathcal{H}^2 \llcorner \hat{\Sigma}^\infty, 0) = \theta^2(\mathcal{H}^2 \llcorner \Sigma^\infty, \infty) < 2. \tag{5.4.2}$$

Clearly, $0 \in \overline{\hat{\Sigma}^\infty}$ if and only if Σ^∞ is non-compact. By smooth convergence $\hat{f}^j := I \circ f^j \to \hat{f}^\infty$ in compact subsets of $\mathbb{R}^3 - \{0\}$ and (1.1.9), we get

$$\int_{\hat{\Sigma}^\infty} |A_{\hat{\Sigma}^\infty}|^2 \, d\mathcal{H}^2 \le \liminf_{j\to\infty} \int_{\Sigma} |A_{\hat{f}^j}|^2 \, d\mu_{\hat{f}_j}$$

$$= \liminf_{j\to\infty} 4\mathcal{W}(\hat{f}^j) - 4\pi\chi(\Sigma) \le 4\pi(8 - \chi(\Sigma)) < \infty.$$

By (5.4.2), smoothness of $\overline{\hat{\Sigma}^\infty}$ follows from Lemma 5.3.2.

Finally, if $\overline{\hat{\Sigma}^\infty}$ is a sphere, it is a Willmore sphere with energy

$$\mathcal{W}(\overline{\hat{\Sigma}^\infty}) \le \liminf_{j\to\infty} \mathcal{W}(\hat{f}^j) \le 8\pi - \delta,$$

hence by Bryant's result (5.4.1) in [Bry82] it has to be a round sphere. Then Σ^∞ is a plane or a round sphere. As we have assumed that Σ^∞ is not compact, it has to be a plane, and f^∞ is trivial. □

Theorem 5.4.1 ([KuSch04, Theorem 5.2]). *Let $f_t : S^2 \to \mathbb{R}^3$ be the Willmore flow of a sphere with*

$$\mathcal{W}(f_0) \le 8\pi. \tag{5.4.3}$$

Then f_t exists globally in time and converges for $t \to \infty$ to a round sphere.

Proof. If $\mathcal{W}(f_0) = 8\pi$, then f_0 is not Willmore according to Bryant's result (5.4.1) in [Bry82]. Therefore the flow is not stationary, hence $\mathcal{W}(f_t) < 8\pi$ for $t > 0$ and when considering the flow $f_{\tau+t}$ for some $\tau > 0$, we may assume that $\mathcal{W}(f_t) \le 8\pi - \delta$ for some $\delta > 0$.

By Proposition 3.A.1 there exists a non-trivial limit under the blow up procedure

$$f^j := r_{t_j}^{-1}(f_{t_j+c_0\varrho_{t_j}^4} - y_{t_j}) \to f^\infty \quad \text{smoothly on compact subsets of } \mathbb{R}^3.$$

When this limit is non-compact, then by the previous Proposition 5.4.1, appropriate inversions $\hat{\Sigma}^\infty = I(f^\infty(\Sigma^\infty)) \cup \{0\}$ are smooth closed Willmore surfaces, which are no spheres. On the other hand by elementary arguments, one has the lower semi-continuity of the genus

$$\text{genus}(\hat{\Sigma}^\infty) \le \liminf_{j\to\infty} \text{genus} f_t(S^2) = 0,$$

which is a contradiction.

Therefore the limit f^∞ is compact. Then the convergence $f^j \to f^\infty$ is smooth, hence $\Sigma^\infty \cong S^2$ and $f^\infty(\Sigma^\infty)$ is a smooth Willmore sphere with energy less than 8π. By Bryant's result (5.4.1) in [Bry82], it has to be a round sphere, and we conclude further by smooth convergence that

$$\mathcal{W}(f_{t_j+c_0\varrho_{t_j}^4}) = \mathcal{W}(f^j) \to \mathcal{W}(f^\infty) = 4\pi,$$

hence by Theorem 3.4.1 the flow f_t exists globally in time and converges for $t \to \infty$ to a round sphere. □

Remarks 5.2.

1. As already remarked in Section 3.3, blow ups and blow downs are never compact by [ChFaSch09]. The last argument in the above proof excludes even without this result that blow ups or blow downs are compact spheres. Therefore supplementing Proposition 5.4.1 we have that blow ups and blow downs are never spheres in codimension one.
2. In codimension 2, Montiel has proved in [Mo00] that Willmore spheres $M^2 \subseteq \mathbb{R}^4$ which are not round spheres satisfy

$$\mathcal{W}(M^2) \ge 8\pi.$$

Combining this result with Riviere's result in [Ri08] the above applications are true in $\mathbb{R}^4$, when in Proposition 5.4.1 and in Theorem 5.4.1 (5.4.3) the strict inequality

$$\mathcal{W}(f_0) < 8\pi$$

is assumed.

References

[Bry82] R. BRYANT, *A duality theorem for Willmore surfaces*, Journal of Differential Geometry, **20** (1984), 23–53.

[Es88] J. H. ESCHENBURG, *Willmore surfaces and Moebius Geometry*, manuscript (1988).

[FaKr] H. M. FARKAS and I. KRA, "Riemann Surfaces", Springer Verlag, Berlin - Heidelberg - New York, 1991.

[GT] D. GILBARG and N. S. TRUDINGER, "Elliptic Partial Differential Equations of Second Order", Springer Verlag, 3.Auflage, Berlin - Heidelberg - New York - Tokyo, 1998.

[Hu57] A. HUBER, *On subharmonic functions and differential geometry in the large*, Comment. Math. Helvetici, **32** (1957), 13–72.

[KuSch01] E. KUWERT and R. SCHÄTZLE, *The Willmore Flow with small initial energy*, Journal of Differential Geometry, **57** (2001), 409–441.

[KuSch02] E. KUWERT and R. SCHÄTZLE, *Gradient flow for the Willmore functional*, Communications in Analysis and Geometry, **10** (2002), 307–339.

[KuSch04] E. KUWERT and R. SCHÄTZLE, *Removability of point singularities of Willmore surfaces*, Annals of Mathematics, **160** (2004), 315–357.

[Mo00] S. MONTIEL,*Spherical Willmore surfaces in the four-sphere*, Transactions of the American Mathematical Society, **352** (2000), 4469–4486.

[MuSv95] S. MÜLLER and V. ŠVERÁK, *On surfaces of finite total curvature*, Journal of Differential Geometry, **42** (1995), 229–258.

[Os69] R. OSSERMAN, "A Survey of Minimal Surfaces", Van Nostrand, 1969.

[Ri08] T. RIVIÈRE, *Analysis aspects of Willmore surfaces*, Inventiones Mathematicae, **174** (2008), 1–45.

[Sim96] L. SIMON, "Singular Sets and Asymptotics in Geometric Analysis", Lipschitz lectures, Institut für Angewandte Mathematik, Universität Bonn, 1996.

6 Compactness modulo the Möbius group

6.1 Non-compact invariance group

As we have seen in Proposition 1.2.3, the Willmore functional is invariant under conformal changes of the ambient space, that is under the full Möbius group, hence has a non-compact invariance group. In particular this means, the Willmore functional cannot be minimized by the direct method of calculus of variations. For if f were already a minimizer, a global one or under certain conformal invariant constraints, e.g. under fixed genus, then for a non-compact sequence of Möbius transformations Φ_j the sequence $\Phi_j \circ f$ were a minimizing, but non-compact sequence.

In this part we address the question of compactness after dividing out the Möbius group. We consider the following approach. Let $f : \Sigma \to \mathbb{R}^n$ be a smooth immersion of a closed, orientable surface Σ and $g = f^* g_{\text{euc}}$ the pull-back metric of the Euclidean metric under f. Then by Poincaré's theorem or the uniformisation theorem for simply connected Riemann surfaces, see [FaKr, Theorem IV.4.1], there exists a metric $g_{\text{poin}} := e^{-2u} g$ on Σ conformal to g with constant curvature $K_{g_{\text{poin}}} \equiv \text{const}$ and unit volume $\text{vol}(g_{\text{poin}}) = 1$. If $\Sigma \not\approx S^2$ or likewise $\text{genus}(\Sigma) \geq 1$, then g_{poin} and u are unique. Our aim is to estimate u, and again this is not possible in presence of the non-compact invariance group.

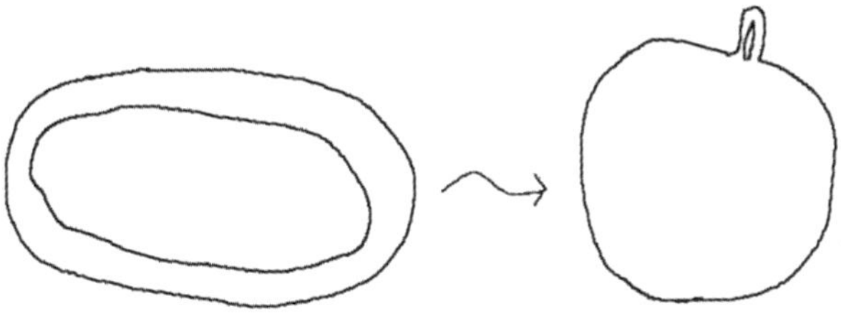

In the above picture, we deform a torus via a Möbius transformation Φ in a sphere with a small handle. This does not change the Willmore energy, and the new pull-back metric $\tilde{g} := (\Phi \circ f)^* g_{\text{euc}} = f^* \Phi^* g_{\text{euc}}$ is conformal to $g = f^* g_{\text{euc}}$, as the Möbius transformation is a conformal transformation of Euclidean space $\mathbb{R}^n$. Therefore $\tilde{g}$ is conformal to $g_{\text{poin}} = e^{-2\tilde{u}} \tilde{g}$. When the handle is getting smaller, the image approaches a sphere, and the conformal factor $\tilde{u}$ degenerates.

The following theorem is the main result in this section.

Theorem 6.1.1 ([KuSch11, Theorem 4.1]). *Let $f : T^2 \to \mathbb{R}^3$ be a smooth immersion of a torus with Willmore energy*

$$\mathcal{W}(f) \leq 8\pi - \delta.$$

Then there exists a Möbius transformation Φ *such that*

$$(\Phi \circ f)^* g_{\text{euc}} = e^{2u} g_{\text{poin}}$$

with g_{poin} *of constant curvature and unit volume and*

$$\| u \|_{L^\infty(\Sigma)} \leq C_\delta.$$

Remarks 6.1.

1. More precisely [KuSch11, Theorem 4.1] states for a smooth immersion $f : \Sigma \to \mathbb{R}^n, n = 3, 4,$ of a closed, orientable surface Σ of genus $p \geq 1$ and with Willmore energy

$$\mathcal{W}(f) \leq \mathcal{W}_{n,p} - \delta,$$

where

$$\mathcal{W}_{n,p} := \min \begin{pmatrix} 8\pi, & \\ 4\pi + \sum_k (\beta^n_{p_k} - 4\pi) \text{ for } \sum_k p_k = p, 1 \leq p_k < p, \\ \beta^4_p + \frac{8\pi}{3} & \text{for } n = 4, \end{pmatrix}$$

and

$$\beta^n_p := \inf\{W(f) \mid f : \Sigma \to \mathbb{R}^n, \text{genus}(\Sigma) = p, \Sigma \text{ orientable }\},$$

there exists a Möbius transformation Φ such that

$$(\Phi \circ f)^* g_{\text{euc}} = e^{2u} g_{\text{poin}}$$

with g_{poin} constant curvature and unit volume and

$$\| u \|_{L^\infty(\Sigma)} \leq C_{p,\delta}.$$

2. 8π in the above minimium is necessary. To this end, we consider two concentric spheres

and connect them by $p + 1$ necks. This gives an orientable surface of genus p and with Willmore energy $8\pi + \varepsilon$. Making ε arbitrarily small and letting the surface converge to a double sphere, the conformal factor degenerates even after applying Möbius tranformations, as Möbius transformations can enlarge at most one neck.

3. In [BaKu03] the strict inequality

$$\beta_p^n - 4\pi < \sum_k (\beta_{p_k}^n - 4\pi) \quad \text{for} \sum_k p_k = p,\, 1 \le p_k < p$$

was proved, which shows that the second condition above is non-empty. On the other hand, glueing together orientable surfaces of genus p_k which nearly minimize the Willmore energy to an orientable surface of genus $p = \sum_k p_k$ appropriately shows that without the second condition above the conformal factor degenerates even after applying Möbius tranformations.

4. If $\beta_q^n \ge 6\pi$ for $1 \le q < p, n = 3$ and $\le p, n = 4$, then $\mathcal{W}_{n,p} = 8\pi$. In particular $\mathcal{W}_{3,1} = 8\pi$. We observe the Willmore conjecture states $\mathcal{W}(T^2) \ge 2\pi^2 > 6\pi$.

5. Combining the above theorem with Mumford's compactness theorem, see [Tr, Theorem C.1], one obtains that the conformal class induced by immersions $f : \Sigma \to \mathbb{R}^n, n = 3, 4, \text{genus}(\Sigma) = p \ge 1, \Sigma$ orientable and Willmore energy $W(f) \le \mathcal{W}_{n,p} - \delta$ are compactly contained in the moduli space.
As a conclusion we obtain that a torus $T^2 \subseteq \mathbb{R}^3$ which is conformally equivalent to a quotient $\mathbb{C}/(\mathbb{Z} + \omega\mathbb{Z}), \omega = x + iy \notin \mathbb{R},$ and $y \ge y_\delta$, satisfies

$$\mathcal{W}(T^2) \ge 8\pi - \delta,$$

in particular cannot compete for the Willmore conjecture.

6. For an immersion $f : \Sigma \to \mathbb{R}^n, n = 3, 4$ of a closed, orientable surface Σ of genus $p \ge 1$, conformal to a given smooth metric g_0 on Σ and with Willmore energy $\mathcal{W}(f) \le \mathcal{W}_{n,p} - \delta$, there exists a Möbius transformation Φ such that

$$(\Phi \circ f)^* g_{\text{euc}} = e^{2u} g_{\text{poin}}$$

with g_{poin} constant curvature and unit volume and

$$\| u \|_{L^\infty(\Sigma)} \le C_{p,\delta}.$$

Obviously $g_{\text{poin}} \sim g_0$ are conformal and therefore g_{poin} is uniquely determined by g_0 and independent of f. Writing in local conformal coordinates

$$\Delta_{g_{\text{poin}}}(\Phi \circ f) = e^{2u} \Delta_g(\Phi \circ f) = e^{2u} \vec{\mathbf{H}}_f,$$

where the right-hand side is uniformally bounded in L^2, as $\mathcal{W}_{n,p} \leq 8\pi$, we get after an appropriate translation

$$\parallel \Phi \circ f \parallel_{W^{2,2}(\Sigma)} \leq C(p, \delta, g_0),$$

and moreover

$$c_{0,p,\delta} \leq \text{area}\,(\Phi \circ f) \leq C_{p,\delta}.$$

6.2 Selection of the Möbius transformation

In this section, we specify the selection of the Möbius transformation in Theorem 6.1.1.

Lemma 6.2.1. *Let* $f : \Sigma \to \mathbb{R}^n$ *be an immersion of a closed surface with*

$$E := \int_\Sigma |A^0|^2 \, \mathrm{d}\mu_f \leq E_0$$

Then there exists a Möbius transformation Φ *such that*

$$(\Phi \circ f)(\Sigma) \subseteq B_1(0),$$
$$\int_{B_{\varrho_0}} |A^0|^2 \, \mathrm{d}\mu_f \leq E/2 \quad \textit{for all } B_{\varrho_0} \subseteq \mathbb{R}^n$$

where $\varrho_0 = \varrho_0(n, E_0) > 0$.

Proof. Clearly, we may assume $E > 0$. After translation and rescaling, we may assume

$$\begin{gathered} \int_{B_1} |A^0|^2 \, \mathrm{d}\mu_f \leq E/2 \quad \text{for all } B_1 \subseteq \mathbb{R}^n, \\ \int_{\overline{B_1(0)}} |A^0|^2 \, \mathrm{d}\mu_f \geq E/2. \end{gathered} \tag{6.2.1}$$

Our aim is to select a point $x_0 \in \mathbb{R}^n$ satisfying

$$\begin{gathered} \overline{B_1(x_0)} \cap f(\Sigma) = \emptyset, \\ |x_0| \leq C(n, E_0). \end{gathered} \tag{6.2.2}$$

By Gauß-Bonnet's theorem in (1.1.9)

$$\int_\Sigma |A|^2 \, \mathrm{d}\mu_f, \mathcal{W}(f) \leq C(E_0),$$

hence by (2.1.15)

$$\varrho^{-2}\mu_f(B_\varrho) \le C\mathcal{W}(f) \le C(E_0) \quad \forall B_\varrho \subseteq \mathbb{R}^n$$

and by Proposition 2.1.2 (2.1.20)

$$\mu_f(B_1(x)) \ge c_0(E_0) > 0 \quad \forall x \in f(\Sigma),$$

when observing that $f(\Sigma) \not\subseteq B_1$, as $\int_{B_1} |A^0|^2 \, \mathrm{d}\mu_f \le E/2 < E = \int_\Sigma |A^0|^2 \, \mathrm{d}\mu_f$ by (6.2.1).

For $R < \infty$ large chosen below, we select a maximal pairwise disjoint family of balls $B_2(x_j) \subseteq B_R(0)$, $j = 1, \dots, N$. Clearly $B_R(0) \subseteq \cup_{j=1}^N B_4(x_j)$, hence $N \ge R^n/4^n$. As $\mu_f(B_R(0)) \le C(E_0)R^2$ from above and $n > 2$, we can choose $R = R(n, E_0)$ large such that $N \ge R^n/4^n > C(E_0)R^2/c_0(E_0)$, hence

$$\mu_f(B_2(x_j)) < c_0(E_0) \quad \text{for some } j.$$

Then clearly $\overline{B_1(x_j)} \cap f(\Sigma) = \emptyset$, $|x_j| \le R = R(n, E_0)$, and we obtain (6.2.2) for $x_0 := x_j$.

Translating f by $-x_0$, we see

$$\overline{B_1(0)} \cap f(\Sigma) = \emptyset,$$
$$\int\limits_{\overline{B_1(-x_0)}} |A^0|^2 \, \mathrm{d}\mu_f \ge E/2,$$

hence, as $|x_0| \le C(n, E_0)$,

$$\int\limits_{\mathbb{R}^n - B_{C(n,E_0)}(0)} |A^0|^2 \, \mathrm{d}\mu_f \le E/2.$$

Moreover we keep by (6.2.1) that

$$\int\limits_{B_1} |A^0|^2 \, \mathrm{d}\mu_f \le E/2 \quad \text{for all } B_1 \subseteq \mathbb{R}^n.$$

Now we invert at the unit sphere that is we consider the Möbius transformation $\Phi(x) := x/|x|^2$ and put $\tilde{f} := \Phi \circ f$. Clearly $\tilde{f}(\Sigma) = \Phi(F(\Sigma)) \subseteq B_1(0)$. By conformal invariance of the Willmore functional in Proposition 1.2.1 (1.2.3), we see for any Borel set $B \subseteq \mathbb{R}^n \cup \{\infty\}$

$$\int\limits_B |A_f^0|^2 \, \mathrm{d}\mu_f = \int\limits_{\Phi(B)} |A_{\tilde{f}}^0|^2 \, \mathrm{d}\mu_{\tilde{f}}.$$

Putting $r := 1/(2C(n, E_0))$, we see for any ball B_ϱ with $B_\varrho \cap B_r(0) \neq \emptyset, \varrho \leq r$ that

$$\int_{B_\varrho} |A^0_{\tilde{f}}|^2 \, d\mu_{\tilde{f}} \leq \int_{B_{2r}(0)} |A^0_{\tilde{f}}|^2 \, d\mu_{\tilde{f}} = \int_{\mathbb{R}^n - B_{C(n,E_0)}(0)} |A^0_f|^2 \, d\mu_f \leq E/2.$$

On the other hand, we see for $B_\varrho \cap B_r(0) = \emptyset$ that $|\nabla \Phi| \leq r^{-2}$ on B_ϱ, hence $\Phi(B_\varrho) \subseteq \tilde{B}_{\varrho r^{-2}}$ and for $\varrho \leq r^2$

$$\int_{B_\varrho} |A^0_{\tilde{f}}|^2 \, d\mu_{\tilde{f}} \leq \int_{\Phi(B_\varrho)} |A^0_f|^2 \, d\mu_f \leq \int_{\tilde{B}_1} |A^0_f|^2 \, d\mu_f \leq E/2.$$

Now the conculsion follows for $\varrho_0(n, E_0) := r^2 \leq 1$. □

6.3 Proof of the compactness theorem

Here we give a proof of Theorem 6.1.1:
We consider an embedded torus $T^2 \subseteq \mathbb{R}^3$ with Willmore energy $\mathcal{W}(T^2) \leq 8\pi - \delta$ as in Theorem 6.1.1. We can assume embeddedness by Proposition 2.1.1. By Gauß-Bonnet's theorem in (1.1.9)

$$\int_{T^2} |A^0_{T^2}|^2 \, d\mathcal{H}^2 = \frac{1}{2} \int_{T^2} |A_{T^2}|^2 \, d\mathcal{H}^2 = 2\mathcal{W}(T^2) \leq 16\pi.$$

Thus by the selection principle Lemma 6.2.1, we can assume after applying a suitable Möbius transformation that

$$\begin{aligned} T^2 &\subseteq B_1(0), \\ \int_{B_{\varrho_0}} |A^0_{T^2}|^2 \, d\mathcal{H}^2 &\leq \frac{1}{2} \int_{T^2} |A^0_{T^2}|^2 \, d\mathcal{H}^2 = \mathcal{W}(T^2) \\ &\leq 8\pi - \delta \quad \forall B_{\varrho_0} \subseteq \mathbb{R}^3 \end{aligned} \tag{6.3.1}$$

where $\varrho_0 > 0$. We cover

$$T^2 \subseteq \cup_{k=1}^K B_{\varrho/4}(x_k)$$

with

$$\int_{B_\varrho(x_k) - B_{c_0(\delta)\varrho}(x_k)} |A|^2 \, d\mathcal{H}^2 \llcorner T^2 \leq \varepsilon^2 \tag{6.3.2}$$

for ε small enough, and we may assume that $\varrho \leq \varrho_0$. Actually, we had to consider a sequence and do this cover uniformly that $K \leq \Lambda, \varrho \geq \Lambda^{-1}$

and then estimate the exponent of the conformal factor in terms of Λ, which we will do here for a single T^2.

By the graphical decomposition lemma [Sim, Lemma 2.1] for ε small enough there exists $\varrho/2 < \sigma_k < 3\varrho/4$ such that

$$T^2 \cap \partial B_{\sigma_k}(x_k) \text{ is a single closed curve.}$$

This is only one curve, since $\mathcal{W}(T^2) \leq 8\pi - \delta$ and choosing $c_0(\delta)$ small enough, when adapting the Li-Yau inequality (2.1.17). We further claim

$$T^2 \cap B_{\sigma_k}(x_k) \text{ is a disc.} \tag{6.3.3}$$

Indeed by Gauß-Bonnet's theorem

$$-2\pi\chi(T^2 \cap B_{\sigma_k}(x_k)) = -\int\limits_{T^2 \cap B_{\sigma_k}(x_k)} K \, \mathrm{d}\mathcal{H}^2 - \int\limits_{T^2 \cap \partial B_{\sigma_k}(x_k)} \kappa_g \, \mathrm{d}\mathcal{H}^1,$$

where κ_g denots the geodesic curvature.

Actually by the graphical decomposition lemma [Sim, Lemma 2.1] the radius σ_k can be choosen such that $T^2 \cap \partial B_{\sigma_k}(x_k)$ is close to a circle, in particular

$$\int\limits_{T^2 \cap \partial B_{\sigma_k}(x_k)} \kappa_g \, \mathrm{d}\mathcal{H}^1 = 2\pi + O(\varepsilon^2). \tag{6.3.4}$$

Observing $-K \leq |A^0|^2/2$ by (1.1.6), we continue with (6.3.1)

$$\begin{aligned} -2\pi\chi(T^2 \cap B_{\sigma_k}(x_k)) &\leq \frac{1}{2} \int\limits_{T^2 \cap B_{\varrho_0}(x_k)} |A^0|^2 \, \mathrm{d}\mathcal{H}^2 - 2\pi + C\varepsilon^2 \\ &\leq 2\pi - \delta/2 + C\varepsilon^2 < 2\pi \end{aligned}$$

for $C\varepsilon^2 < \delta/2$, hence

$$\chi(T^2 \cap B_{\sigma_k}(x_k)) > -1,$$

hence $\chi(T^2 \cap B_{\sigma_k}(x_k)) = 1$, and we get (6.3.3).

Then by Gauß-Bonnet's theorem and (6.3.4)

$$\int\limits_{T^2 \cap B_{\sigma_k}(x_k)} K \, \mathrm{d}\mathcal{H}^2 = 2\pi\chi(T^2 \cap B_{\sigma_k}(x_k)) - \int\limits_{T^2 \cap \partial B_{\sigma_k}(x_k)} \kappa_g \, \mathrm{d}\mathcal{H}^1 = O(\varepsilon^2).$$

By (1.1.2) and (1.1.7), we get by (6.3.1)

$$\begin{aligned}\int\limits_{T^2\cap B_{\sigma_k}(x_k)} |K| \, d\mathcal{H}^2 &\leq \frac{1}{2} \int\limits_{T^2\cap B_{\sigma_k}(x_k)} |A|^2 \, d\mathcal{H}^2 \\ &= \int\limits_{T^2\cap B_{\sigma_k}(x_k)} |A^0|^2 \, d\mathcal{H}^2 + \int\limits_{T^2\cap B_{\sigma_k}(x_k)} K \, d\mathcal{H}^2 \\ &\leq 8\pi - \delta + C\varepsilon^2 \leq 8\pi - \delta/2\end{aligned}$$

for $C\varepsilon^2 < \delta/2$.

By (6.3.1) and since $T^2 \cap \partial B_{\sigma_k}(x_k)$ is close to a circle, we can extend $T^2 \cap B_{\sigma_k}(x_k)$ to a plane $\Sigma_k \subseteq \mathbb{R}^3, \cong \mathbb{R}^2$, such that

$$\begin{aligned}&\Sigma_k \cap B_{\varrho/2}(x_k) = T^2 \cap B_{\varrho/2}(x_k), \\ &\Sigma_k - B_\varrho(x_k) \text{ is a plane minus a ball}, \\ &\int\limits_{\Sigma_k} |K| \, d\mathcal{H}^2 \leq 8\pi - \delta/4, \\ &\int\limits_{\Sigma_k} |A|^2 \, d\mathcal{H}^2 \leq 16\pi.\end{aligned}$$

Then by uniformisation theorem for simply connected Riemann surfaces, see [FaKr, Theorem IV.4.1], or more precisely by Huber's theorem, see [Hu57] or Theorem 4.C.1, there is a conformal diffeomorphism $f_k : \mathbb{R}^2 \xrightarrow{\approx} \Sigma_k$, say $g_k := g_{\text{euc}}|\Sigma_k$ and $f_k^*(e^{-2u_k} g_k) = g_{\text{euc}}$, that is $e^{-2u_k} g_k$ is isometric to an Euclidean metric on $\mathbb{R}^2$. By Theorem 4.1.1 after subtracting a suitable constant

$$\begin{aligned}&- \Delta_{g_k} u_k = K_{g_k} \quad \text{on } \Sigma_k, \\ &\| u_k \|_{L^\infty(\Sigma_k)}, \int\limits_{\Sigma_k} |Du_k|_{g_k}^2 \, d\mu_{g_k} \leq C_\delta \int\limits_{\Sigma_k} |A|^2 \, d\mathcal{H}^2 \leq C_\delta.\end{aligned}$$

Writing $g = g_{\text{euc}}|T^2$, we see $g = g_k$ in $T^2 \cap B_{\varrho/2}(x_k)$ and

$$\begin{aligned}&- \Delta_g u_k = K_g \quad \text{on } T^2 \cap B_{\varrho/2}(x_k), \\ &\| u_k \|_{L^\infty(T^2\cap B_{\varrho/2}(x_k))}, \int\limits_{T^2\cap B_{\varrho/2}(x_k)} |Du_k|_g^2 \, d\mu_g \leq C_\delta.\end{aligned}$$

Now we write by Poincaré's theorem $g = e^{2u} g_{\text{poin}}$ with g_{poin} of constant curvature and unit volume. As we are on a torus, we get by Gauß-Bonnet's theorem that $K_{g_{\text{poin}}} \equiv 0$ and $-\Delta_g u = K_g$ by elementary differential geometry. We calculate for $\lambda \in u(T^2)$

$$\int_{T^2} |Du|_g^2 \, \mathrm{d}\mu_g = \int_{T^2} K_g(u - \lambda) \, \mathrm{d}\mu_g \leq \frac{1}{2} \int_{T^2} |A_{T^2}|^2 \, \mathrm{d}\mathcal{H}^2 \operatorname{osc} u$$
$$= 2\mathcal{W}(T^2) \operatorname{osc} u \leq 16\pi \operatorname{osc} u.$$

From above we have seen that $e^{-2u_k} g = e^{-2u_k} g_k$ is isometric to an Euclidean metric in $T^2 \cap B_{\varrho/2}(x_k)$. As $|u_k| \leq C_\delta$, we see for $c_0(\delta)$ small and any $z \in T^2 \cap B_{3\varrho/8}(x_k) \subseteq \Sigma_k$ that the respective Euclidean Ball $B^2_{c_0(\delta)\varrho}(z) \subseteq T^2 \cap B_{\varrho/2}(x_k)$. By Poincaré's inequaliy there exists $\lambda_z \in \mathbb{R}$ such that

$$(c_0\varrho)^{-1} \parallel u - \lambda_z \parallel_{L^2(B_{c_0\varrho}(z))} \leq C \parallel Du \parallel_{L^2(B_{c_0\varrho}(z))}$$
$$\leq C_\delta \parallel |Du|_g \parallel_{L^2(T^2,\mu_g)} \leq C_\delta \sqrt{\operatorname{osc} u}.$$

By graphical decomposition lemma [Sim, Lemma 2.1] there exists $5\varrho/16 < \sigma'_k < 3\varrho/8$ such that $T^2 \cap B_{\sigma'_k}(x_k)$ is a disc. Now we select a maximal subset $z_1, \ldots, z_J \in T^2 \cap B_{\sigma'_k}(x_k)$ with $B^2_{c_0\varrho/4}(z_i) \cap B^2_{c_0\varrho/4}(z_j) = \emptyset$ for $i \neq j$. Clearly $T^2 \cap B_{\sigma'_k}(x_k) \subseteq \cup_{j=1}^J B^2_{c_0\varrho/2}(z_j)$, hence by Proposition 2.1.2 (2.1.20)

$$\mathcal{H}^2(T^2 \cap B^2_{c_0\varrho/2}(z_j)) \geq \mathcal{H}^2(T^2 \cap B_{c_1\varrho}(z_j)) \geq c_0 c_1^2 \varrho^2 / \mathcal{W}(T^2) \geq c_0 \varrho^2,$$

by (2.1.15) and pairwise disjointness

$$c_0 J \varrho^2 \leq \mathcal{H}^2(\cup_{j=1}^J T^2 \cap B^2_{c_0\varrho/2}(z_j)) \leq \mathcal{H}^2(T^2 \cap B_{\varrho/2}(x_k)) \leq C\mathcal{W}(T^2)\varrho^2$$

and

$$J \leq C_\delta.$$

Now if $B^2_{c_0\varrho/2}(z_i) \cap B^2_{c_0\varrho/2}(z_j) \neq \emptyset$, then

$$\mathcal{L}^2(B^2_{c_0\varrho}(z_i) \cap B^2_{c_0\varrho}(z_j)) \geq c_0 \pi \varrho^2,$$

and we calculate from above

$$|\lambda_{z_i} - \lambda_{z_j}| \leq (c_0\varrho)^{-1} \Big(\parallel u - \lambda_{z_i} \parallel_{L^2(B_{c_0\varrho}(z_i))} + \parallel u - \lambda_{z_j} \parallel_{L^2(B_{c_0\varrho}(z_j))} \Big)$$
$$\leq C_\delta \sqrt{\operatorname{osc} u}.$$

As $T^2 \cap B^2_{\sigma'_k}(x_k)$ is connected and covered by the $B^2_{c_0\varrho/2}(z_i)$, we find for i, j a chain $B^2_{c_0\varrho/2}(z_{i_\nu})$, $\nu = 1, \dots, N$ with $N \leq J \leq C_\delta$ and such that neighbouring discs intersect. Thus

$$|\lambda_{z_i} - \lambda_{z_j}| \leq C_\delta \sqrt{\operatorname{osc} u} \quad \forall i, j.$$

Therefore there exists $\lambda_k \in \mathbb{R}$ such that

$$\varrho^{-1} \parallel u - \lambda_k \parallel_{L^2(T^2 \cap B_{5\varrho/16}(x_k))} \leq C_\delta \sqrt{\operatorname{osc} u}.$$

Now if $z \in T^2 \cap B_{\varrho/4}(x_k) \cap B_{\varrho/4}(x_l) \neq \emptyset$, then $B_{\varrho/16}(z) \subseteq B_{5\varrho/16}(x_k) \cap B_{5\varrho/16}(x_l)$ and by Proposition 2.1.2 (2.1.20)

$$\mathcal{H}^2(T^2 \cap B_{\varrho/4}(x_k) \cap B_{\varrho/4}(x_l)) \geq \mathcal{H}^2(T^2 \cap B_{\varrho/16}(z)) \geq c_0 \varrho^2$$

and

$$\begin{aligned} |\lambda_k - \lambda_l| \leq & \varrho^{-1}\Big(\parallel u - \lambda_k \parallel_{L^2(T^2 \cap B_{5\varrho/16}(x_k))} + \parallel u - \lambda_l \parallel_{L^2(T^2 \cap B_{5\varrho/16}(x_l))}\Big) \\ \leq & C_\delta \sqrt{\operatorname{osc} u}. \end{aligned}$$

By connectedness of $T^2 \subseteq \cup_{k=1}^K B_{\varrho/4}(x_k)$, we see

$$|\lambda_k - \lambda_l| \leq C_\delta K \sqrt{\operatorname{osc} u} \quad \forall k, l,$$

and there exists $\lambda \in \mathbb{R}$ such that

$$\begin{aligned} \varrho^{-1} \parallel u - \lambda \parallel_{L^2(T^2)} \leq & \sum_{k=1}^{K} \Big(\varrho^{-1} \parallel u - \lambda_k \parallel_{L^2(T^2 \cap B_{5\varrho/16}(x_k))} + C|\lambda_k - \lambda| \Big) \\ \leq & C_\delta K \sqrt{\operatorname{osc} u}. \end{aligned}$$

Recalling $\varrho \geq \Lambda^{-1}$, $K \leq \Lambda$, we get

$$\parallel u - \lambda \parallel_{L^2(T^2)} \leq C(\Lambda, \delta) \sqrt{\operatorname{osc} u}.$$

From above we see that $u - u_k$ is harmonic with respect to g in $T^2 \cap B_{\varrho/2}(x_k)$, hence also harmonic with respect to Euclidean metric $e^{-2u_k} g$. We choose $z_0 \in T^2$ with $u(z_0) = \min u =: m$. Then $z_0 \in T^2 \cap B_{\varrho/4}(x_k)$ for some k and as above we consider the Euclidean ball $B^2_{c_0\varrho(z_0)} \subseteq T^2 \cap B_{\varrho/2}(x_k)$. Clearly $u - u_k - m + C_\delta \geq 0$ in $T^2 \cap B_{\varrho/2}(x_k)$ and by Harnack inequality

$$\sup_{B^2 c_0 \varrho(z_0)} (u - u_k - m + C_\delta) \leq C \inf_{B^2 c_0 \varrho(z_0)} (u - u_k - m + C_\delta) \leq C_\delta$$

and

$$u - m \le C_\delta \quad \text{in } B^2_{c_0\varrho}(z_0).$$

As $\mathcal{L}^2(B^2_{c_0\varrho}(z_0)) \ge c_0\varrho^2$, we get

$$\begin{aligned}\lambda - m &\le C\varrho^{-1} \parallel \lambda - u + C_\delta \parallel_{L^2(B^2_{c_0\varrho}(z_0))} \\ &\le C\Lambda \parallel u - \lambda \parallel_{L^2(T^2)} + C_\delta \le C(\Lambda, \delta)(1 + \sqrt{\operatorname{osc} u}).\end{aligned}$$

Likewise, we obtain for $M := \max u$ that

$$M - \lambda \le C(\Lambda, \delta)(1 + \sqrt{\operatorname{osc} u}).$$

Putting together yields

$$\operatorname{osc} u = M - m \le C(\Lambda, \delta)(1 + \sqrt{\operatorname{osc} u}),$$

hence $\operatorname{osc} u \le C(\Lambda, \delta)$.

We conclude from (6.3.1) that

$$\varrho_0 \le \operatorname{diam} T^2 \le 2,$$

hence by Proposition 2.1.3 (2.1.21)

$$c_0 \le \mathcal{H}^2(T^2) \le C.$$

As

$$\mathcal{H}^2(T^2) = \operatorname{vol}(g) = \operatorname{vol}(e^{2u} g_{\text{poin}}) = \int_{T^2} e^{2u} \, \mathrm{d}\operatorname{vol}_{g_{\text{poin}}}$$

and $\operatorname{vol}(g_{\text{poin}}) = 1$, we see

$$c_0 \le e^{2M}, e^{2m} \le C,$$

hence $-C \le M, m \le C$. By the estimate above

$$\begin{aligned}M &= m + \operatorname{osc} u \le C + C(\Lambda, \delta) \le C(\Lambda, \delta), \\ m &= M - \operatorname{osc} u \ge -C - C(\Lambda, \delta) \ge -C(\Lambda, \delta),\end{aligned}$$

which finally yields

$$\parallel u \parallel_{L^\infty(T^2)} \le C(\Lambda, \delta),$$

and the theorem is proved. □

References

[BaKu03] M. BAUER and E. KUWERT, *Existence of Minimizing Willmore Surfaces of Prescribed Genus*, IMRN Intern. Math. Res. Notes, **10** (2003), 553–576.

[FaKr] H. M. FARKAS and I. KRA, "Riemann Surfaces", Springer Verlag, Berlin - Heidelberg - New York, 1991.

[KuSch11] E. KUWERT and R. SCHÄTZLE, *Closed surfaces with bounds on their Willmore energy*, to appear in Annali della Scuola Normale Superiore di Pisa, arXiv:math.DG/1009.5286.

[MuSv95] S. MÜLLER and V. ŠVERÁK, *On surfaces of finite total curvature*, Journal of Differential Geometry, **42** (1995), 229–258.

[Tr] A. TROMBA, "Teichmüller Theory in Riemannian Geometry", Birkhäuser, 1992.

[Sim] L. SIMON, "Lectures on Geometric Measure Theory", Proceedings of the Centre for Mathematical Analysis Australian National University, Vol. 3, 1983.

7 Minimizers of the Willmore functional under fixed conformal class

7.1 Constrained Willmore surfaces

In [LY82] respectively in [MoRo86], it was proved that a torus $T^2 \subseteq \mathbb{R}^n$ which is conformally equivalent to a quotient $\mathbb{C}/(\mathbb{Z}+\omega\mathbb{Z})$, $\omega = x+iy \notin \mathbb{R}$, satisfies

$$\mathcal{W}(T^2) \geq \frac{2\pi^2}{y},$$

respectively satisfies

$$\mathcal{W}(T^2) \geq \frac{4\pi^2 y}{1+y^2+x^2-x}.$$

Recalling that the Clifford torus $T_{\text{Cliff}} \cong \mathbb{C}/(\mathbb{Z}+i\mathbb{Z})$ is of square type, we see that any $T^2 \cong T_{\text{Cliff}}$ satisfies

$$\mathcal{W}(T^2) \geq 2\pi^2 = \mathcal{W}(T_{\text{Cliff}}),$$

and the Clifford torus minimizes the Willmore energy in its conformal class. This is the only explicit minimizer of this type known to us.

Considering the Willmore energy for closed, orientable surfaces of fixed conformal class, the Euler Lagrange equation (1.3.10) extends to

$$\Delta^{\perp}\vec{\mathbf{H}} + Q(A^0)\vec{\mathbf{H}} = g^{ik}g^{jl}A^0_{ij}q_{kl} \quad \text{on } \Sigma, \tag{7.1.1}$$

where q is symmetric, traceless, transverse that is

$$\begin{aligned} q_{ij} &= q_{ji}, \\ \operatorname{tr}_g q = g^{ij}q_{ij} &= 0, \\ \operatorname{div}_g q_k = g^{ij}\nabla_i q_{jk} &= 0. \end{aligned}$$

We call immersions into $\mathbb{R}^n$ or S^n which satisfy (7.1.1) for some q constrained Willmore immersions.

We embed the Clifford torus in the family

$$T_r := rS^1 \times \sqrt{1-r^2}S^1 \subseteq S^3 \quad 0 < r \leq 1/\sqrt{2},$$

as clearly $T_{1/\sqrt{2}} = T_{\text{Cliff}}$. These are constant mean curvature surfaces in codimension 1, hence by the next proposition constrained Willmore surfaces.

Proposition 7.1.1. *Let* $f : \Sigma \to \mathbb{R}^3$ *or* S^3 *be a constant mean curvature immersion of a surface* Σ*, that is* $|\vec{\mathbf{H}}_f| \equiv \text{const.}$ *Then* f *is a conformally constrained Willmore immersion.*

Proof. First we consider a constant mean curvature immersion $f : \Sigma \to M$ into any 3-dimensional manifold M. From $|\vec{\mathbf{H}}| \equiv \text{const}$, we see

$$0 = \nabla|\vec{\mathbf{H}}|^2 = 2\langle \nabla^\perp \vec{\mathbf{H}}, \vec{\mathbf{H}} \rangle.$$

If $|\vec{\mathbf{H}}| \neq 0$, then $\vec{\mathbf{H}}$ spans $N\Sigma \subseteq TM$ in codimension 1, and we conclude that the normal derivative vanishes

$$\nabla^\perp \vec{\mathbf{H}} = 0. \tag{7.1.2}$$

In case $|\vec{\mathbf{H}}| \equiv 0$, this is immediate. Then by the equation of Mainardi-Codazzi

$$g^{ij}\nabla_i^\perp A^0_{jk} = g^{ij}\nabla_i^\perp A_{jk} - g^{ij}\nabla_i^\perp\left(\frac{1}{2}g_{jk}\vec{\mathbf{H}}\right) = \frac{1}{2}\nabla_k^\perp \vec{\mathbf{H}} = 0. \tag{7.1.3}$$

Now let $\nu : \Sigma \to N\Sigma \subseteq TM$ locally be a smooth unit normal field at Σ in M. As $|\nu| \equiv 1$, the derivative of $D\nu$ of ν is orthogonal to ν, that is $D\nu \perp \nu$. Since we are in codimension 1, this means that $D\nu$ is tangential at Σ or likewise that the normal derivative $\nabla^\perp \nu = 0$ vanishes.

Defining the scalar mean curvature as $H := \vec{\mathbf{H}}\nu$, we see by (7.1.2)

$$0 = \nabla^\perp \vec{\mathbf{H}} = (\nabla H)\nu,$$

hence $\nabla H \equiv 0$, and the normal laplacian satisfies

$$0 = \Delta_g \vec{\mathbf{H}} = (\Delta_g H)\nu. \tag{7.1.4}$$

Putting $h^0_{ij} := \langle A^0_{ij}, \nu \rangle$, we see likewise by (7.1.3)

$$0 = g^{ij}\nabla_i^\perp A^0_{jk} = (g^{ij}\nabla_i h^0_{jk})\nu,$$

hence $g^{ij}\nabla_i h^0_{jk} \equiv 0$. This means that the divergence of h^0 vanishes, and, as h^0 is tracefree and symmetric, we obtain that h^0 is transverse traceless symmetric with repsect to g. Recalling $\nabla H = 0$, we see that $q_{kl} := \langle A^0_{kl}, \vec{\mathbf{H}} \rangle = h^0_{kl} H$ is transverse traceless symmetric too.

We continue by (7.1.4)

$$\Delta_g \vec{\mathbf{H}} + Q(A^0)\vec{\mathbf{H}} = g^{ik}g^{jl}A^0_{ij}\langle A^0_{kl}, \vec{\mathbf{H}} \rangle = g^{ik}g^{jl}A^0_{ij}q_{kl}.$$

For $M = \mathbb{R}^3$ respectively S^3, this is (7.1.1), and f is a conformally constrained Willmore immersion. □

T_r are of rectangular type

$$T_r \cong \mathbb{C}/(\mathbb{Z} \times i(\sqrt{1-r^2}/r)\mathbb{Z})$$

and clearly

$$\begin{aligned}\mathcal{W}(T_r) &= \frac{1}{4}\mathrm{Area}(T_r)|\vec{\mathbf{H}}_{T_r}|^2 \\ &= \pi^2 r\sqrt{1-r^2}\left(1+\left|\frac{1}{r}-\frac{1}{\sqrt{1-r^2}}\right|^2\right) \sim \frac{1}{r} \to \infty.\end{aligned}$$

On the other hand, the construction in Remark 2. after Theorem 6.1.1 for $p = 1$ and choosing the necks antipodal in order to get a rotational symmetric torus yields a torus a degenerate rectangular type of Willmore energy $8\pi + \varepsilon$. Therefore for small r, the tori T_r do not minimize the Willmore energy in their conformal classes and are only critical points.

7.2 Existence of constrained minimizers

The main goal of this part is the following existence theorem for constrained minimizers.

Theorem 7.2.1 ([KuSch07, Theorem 7.3]). *Let Σ be a closed Riemann surface of genus $p \geq 1$ with conformal metric g_0 and $n = 3, 4$. If*

$$\mathcal{W}(\Sigma, g_0, n) := \inf\{\mathcal{W}(f) \mid f : \Sigma \to \mathbb{R}^n \textit{ immersion conformal to } g_0\} < \mathcal{W}_{n,p},$$

where $\mathcal{W}_{n,p}$ is defined in Remark 1. after Theorem 6.1.1, then there exists a smooth minimizing immersion $f : \Sigma \to \mathbb{R}^n$ conformal to g_0

$$\mathcal{W}(f) = \mathcal{W}(\Sigma, g_0, n).$$

Proof. Firstly to start with the direct method of calculus of variations, we select a minimizing sequence of conformal immersions $f_m : (\Sigma, g_0) \to \mathbb{R}^n$ that is

$$\mathcal{W}(f_m) \to \mathcal{W}(\Sigma, g_0, n).$$

By assumption we may assume for large m that $\mathcal{W}(f_m) < \mathcal{W}_{n,p} - \delta$ for some $\delta > 0$. By Theorem 6.1.1 remark 6. there exist Möbius transformations Φ_m with

$$\begin{aligned}c_{0,p,\delta} g_0 \leq (\Phi_m \circ f_m)^* g_{\mathrm{euc}} \leq C_{p,\delta} g_0, \\ \| \Phi_m \circ f_m \|_{W^{2,2}(\Sigma)} \leq C(p, \delta, g_0).\end{aligned}$$

Clearly $\Phi_m \circ f_m$ is still minimizing and replacing f_m by $\Phi_m \circ f_m$ and passing to a subsequence

$$\begin{aligned} f_m \to f \quad & \text{weakly in } W^{2,2}(\Sigma), \\ c_{0,p,\delta} g_0 \leq f^* & g_{euc} \leq C_{p,\delta} g_0, \end{aligned}$$

and $f_m^* g_{euc} \to f^* g_{euc}$ strongly in $L^2(\Sigma)$, in particular $f^* g_{euc}$ is conformal to g_0.

The remaining question is the smoothness of f. The standard method would be to consider variations $f + tV$ and to do energy comparison. Now $f + tV$ is in general not conformal to g_0 anymore, and the energy comparison is not immediately possible. Even choosing some variations $V_1, \ldots, V_R$ a priori, we cannot in general get a correction $f + tV + \lambda_r V_r$ which is conformal, as being conformal is a pointwise, infinite dimensional condition.

Instead we recall the parameter invariance of the Willmore functional that is

$$\mathcal{W}(f) = \mathcal{W}(f \circ \phi) \quad \forall \phi : \Sigma \xrightarrow{\approx} \Sigma.$$

As we have seen in Section 6 by Poincaré's theorem, any metric on Σ is conformally equivalent to a constant curvature, unit volume metric. Therefore we seek corrections in the moduli space

$$\begin{aligned} & \mathcal{M}_{\text{poin}}/\mathcal{D} \\ & := \{\text{constant curvature, unit volume metrics on} \Sigma\ \}/\{\phi : \Sigma \xrightarrow{\approx} \Sigma\}. \end{aligned}$$

As the moduli space has a bad analytical structure, we divide by the smaller group $\mathcal{D}_0 := \{\phi \in \mathcal{D} \mid \phi \simeq id_\Sigma\ \}$, where $\simeq$ means homotopic, to obtain the Teichmüller space

$$\mathcal{T} := \mathcal{M}_{\text{poin}}/\mathcal{D}_0,$$

see [Tr]. The Teichmüller space is an open finite dimensional manifold of dimension

$$\dim \mathcal{T} = \begin{cases} 2 & \text{for } p = 1, \\ 6p - 6 & \text{for } p \geq 2. \end{cases}$$

We consider the projection $\pi : \mathcal{M} := \{$ metrics on Σ $\} \to \mathcal{T}$ mapping any metric on Σ to the class of the conformally equivalent constant curvature, unit volume metric, which exists and is unique by Poincaré's theorem. Now there is a decomposition of the tangent space of $\mathcal{M}$ into

$$T_g\mathcal{M} := \{\sigma g + \mathcal{L}_X g\} \oplus S_2^{TT}(g),$$

where $S_2^{TT}(g)$ are the 2-covariant tensors on Σ which are symmetric, traceless and divergence free with respect to g. As we have already seen, the elements of $S_2^{TT}(g)$ appear exactly as Lagrange multipliers in the Euler-Lagrange equation (7.1.1) for constrained Willmore immersions. Now we have

$$d\pi_g|\{\sigma g + \mathcal{L}_X g\} = 0,$$
$$d\pi_g|S_2^{TT}(g) \xrightarrow{\approx} T_{\pi g}\mathcal{T},$$
$$\dim\, \mathcal{T} = \dim\, S_2^{TT}(g) = \begin{cases} 2 & \text{for } p = 1, \\ 6p-6 & \text{for } p \geq 2, \end{cases}$$

where the dimension of $S_2^{TT}(g)$ can be calculated by Riemann-Rochs theorem.

For $f : \Sigma \to \mathbb{R}^n$, $g = f^* g_{\text{euc}}$ smooth, actually more precisely for f_m, we define the first Teichmüller variation $\delta\pi_f : C^\infty(\Sigma, \mathbb{R}^n) \to T_{\pi g}\mathcal{T}$ by

$$\delta\pi_f.V := \frac{d}{dt}\pi\Big((f+tV)^* g\Big)|_{t=0}$$

and consider the subspace

$$\mathcal{V}_f := \delta\pi_f C^\infty(\Sigma, \mathbb{R}^n) \subseteq T_{\pi g}\mathcal{T}.$$

We call f of full rank in Teichmüller space if $\mathcal{V}_f = T_{\pi g}\mathcal{T}$, otherwise we call f degenerate.

In the full rank case, we select variations $V_1, \ldots, V_{\dim\,\mathcal{T}}$ with

$$\text{span}\{\delta\pi_f.V_1, \ldots, \delta\pi_f.V_{\dim\,\mathcal{T}}\} = \mathcal{V}_f = T_{\pi g}\mathcal{T}.$$

Then by implicit function theorem for t small, there exists $\lambda \in \mathbb{R}^{\dim\,\mathcal{T}}$ such that

$$\pi\Big((f + tV + \lambda_r V_r)^* g_{\text{euc}}\Big) = \pi(f^* g_{\text{euc}}),$$

hence

$$\mathcal{W}(f) \leq \mathcal{W}(f + tV + \lambda_r V_r)$$

and the standard whole-filling procedure applies to prove $\int_{B_\varrho} |D^2 f|^2\, d\mu \leq C\varrho^{2\alpha}$, $f \in C^{1,\alpha}$, then $f \in C^\infty$ and moreover the Euler-Lagrange equation (7.1.1) for f. This proves the full rank case. □

In the degenerate case, we have $\dim\,\mathcal{V}_f < \dim\,\mathcal{T}$. We first prove that only one dimension can be lost. Writing $g_t := (f+tV)^* g_{\text{euc}}$ and $g = e^{2u} g_{\text{poin}}$ for some constant curvature unit volume metric g_{poin}, we see $\pi(g_t) = \pi(e^{-2u} g_t)$, hence for an orthonormal basis $q^r(g_{\text{poin}}), r = 1, \ldots, \dim \mathcal{T}$, of $S_2^{TT}(g_{\text{poin}})$

$$\begin{aligned}\delta\pi_f.V &= d\pi_g.\partial_t g_{t\,|t=0} = d\pi_{g_{\text{poin}}}.e^{-2u}\partial_t g_{t\,|t=0}\\ &= \sum_{r=1}^{\dim \mathcal{T}} \langle e^{-2u}\partial_t g_{t\,|t=0}.q^r(g_{\text{poin}})\rangle_{g_{\text{poin}}} d\pi_{g_{\text{poin}}}.q^r(g_{\text{poin}}).\end{aligned}$$

Calculating in local charts

$$g_{t,ij} = g_{ij} + t\langle \partial_i f, \partial_j V\rangle + t\langle \partial_j f, \partial_i V\rangle + t^2\langle \partial_i V, \partial_j V\rangle,$$

we get

$$\begin{aligned}\langle e^{-2u}\partial_t g_{t\,|t=0}, q^r(g_{\text{poin}})\rangle_{g_{\text{poin}}} &= \int_\Sigma g_{\text{poin}}^{ik} g_{\text{poin}}^{jl} e^{-2u}\partial_t g_{t,ij\,|t=0} q_{kl}^r(g_{\text{poin}})\, \mathrm{d}\mu_{g_{\text{poin}}}\\ &= 2\int_\Sigma g^{ik} g^{jl}\langle \partial_i f, \partial_j V\rangle q_{kl}^r(g_{\text{poin}})\, \mathrm{d}\mu_g\\ &= -2\int_\Sigma g^{ik} g^{jl}\langle \nabla_j^g \nabla_i^g f, V\rangle q_{kl}^r(g_{\text{poin}})\, \mathrm{d}\mu_g\\ &\quad - 2\int_\Sigma g^{ik} g^{jl}\langle \partial_i f, V\rangle \nabla_j^g q_{kl}^r(g_{\text{poin}})\, \mathrm{d}\mu_g\\ &= -2\int_\Sigma g^{ik} g^{jl}\langle A_{ij}^0, V\rangle q_{kl}^r(g_{\text{poin}})\, \mathrm{d}\mu_g,\end{aligned}$$

as $q \in S_2^{TT}(g_{\text{poin}}) = S_2^{TT}(g)$ by Proposition 7.A.1 is divergence- and tracefree with respect to g. Therefore

$$\delta\pi_f.V = \sum_{r=1}^{\dim \mathcal{T}} 2\int_\Sigma g^{ik} g^{jl}\langle \partial_i f, \partial_j V\rangle q_{kl}^r(g_{\text{poin}})\, \mathrm{d}\mu_g\, d\pi_{g_{\text{poin}}}.q^r(g_{\text{poin}}) =$$

$$= \sum_{r=1}^{\dim \mathcal{T}} -2\int_\Sigma g^{ik} g^{jl}\langle A_{ij}^0, V\rangle q_{kl}^r(g_{\text{poin}})\, \mathrm{d}\mu_g\, d\pi_{g_{\text{poin}}}.q^r(g_{\text{poin}}). \qquad (7.2.1)$$

Proposition 7.2.1. *We always have*

$$\dim \mathcal{V}_f \geq \dim \mathcal{T} - 1. \qquad (7.2.2)$$

f *is not of full rank in Teichmüller space if and only if*

$$g^{ik} g^{jl} A_{ij}^0 q_{kl} = 0 \quad \textit{for some } q \in S_2^{TT}(g_{\text{poin}}), q \neq 0. \qquad (7.2.3)$$

In this case, f is isothermic locally around all but finitely many points of Σ, that is around all but finitely many points of Σ there exist local conformal principal curvature coordinates.

Proof. For $q \in S_2^{TT}(g_{\text{poin}})$, we put $\Lambda_q : C^\infty(\Sigma, \mathbb{R}^n) \to \mathbb{R}$

$$\Lambda_q.V := -2 \int_\Sigma g^{ik} g^{jl} \langle A^0_{ij}, V \rangle q_{kl} \, \mathrm{d}\mu_g$$

and define the annihilator

$$_{\text{Ann}} := \{ q \in S_2^{TT}(g_{\text{poin}}) \mid \Lambda_q = 0 \}.$$

As $d\pi_{g_{\text{poin}}} | S_2^{TT}(g_{\text{poin}}) \to T_{g_{\text{poin}}} \mathcal{T}$ is bijective, we see by (7.2.1) and elementary linear algebra

$$\dim \mathcal{T} = \dim \mathcal{V}_f + \dim {}_{\text{Ann}}. \tag{7.2.4}$$

Clearly,

$$q \in_{\text{Ann}} \Longleftrightarrow g^{ik} g^{jl} A^0_{ij} q_{kl} = 0 \tag{7.2.5}$$

which already yields (7.2.3). In local conformal coordinates $g_{ij} = e^{2u} g_{\text{poin}} = e^{2v} \delta_{ij}$, we see $A^0_{11} = -A^0_{22}, A^0_{12} = A^0_{21}, q_{11} = -q_{22}, q_{12} = q_{21}$, as both A^0 and $q \in S_2^{TT}(g_{\text{poin}})$ are symmetric and tracefree with respect to $g = e^{2u} g_{\text{poin}}$. This rewrites (7.2.5) into

$$q \in_{\text{Ann}} \Longleftrightarrow A^0_{11} q_{11} + A^0_{12} q_{12} = 0. \tag{7.2.6}$$

Now if (7.2.2) were not true, there would be two linearly independent $q^1, q^2 \in_{\text{Ann}}$ by (7.2.4). Likewise the functions $h^k := q^k_{11} - i q^k_{12}$, which are holomorphic by Proposition 7.A.2, are linearly independant over $\mathbb{R}$, in particular neither of them vanishes identically, hence these vanish at most at finitely many points, as Σ is closed. Then h^1/h^2 is meromorphic and moreover not a real constant. This implies that $Im(h^1/h^2)$ does not vanish identically, hence vanishes at most at finitely many points. Outside these finitely many points, we calculate

$$Im(h^1/h^2) = |h^2|^{-2} Im(h^1 \overline{h^2}) = |h^2|^{-2} \det \begin{pmatrix} q^1_{11} & q^1_{12} \\ q^2_{11} & q^2_{12} \end{pmatrix},$$

hence

$$\det \begin{pmatrix} q^1_{11} & q^1_{12} \\ q^2_{11} & q^2_{12} \end{pmatrix} \text{ vanishes at most at finitely many points}$$

and by (7.2.6)

$$A^0 = 0. \tag{7.2.7}$$

Then by a theorem of Codazzi, f parametrizes a round sphere or a plane, contradicting $p \geq 1$, and (7.2.2) is proved.

Next if f is not of full rank in Teichmüller space, there exits $q \in_{\text{Ann}} -\{0\} \neq \emptyset$ by (7.2.4), and the holomorphic function $h = q_{11} - iq_{12}$ vanishes at most at finitely many points. In a neighbourhood of a point where h does not vanish, there is a holomorphic function w with $(w')^2 = ih$. Then w has a local inverse z and using w as new local conformal coordinates, h transforms into $(h \circ z)(z')^2 = -i$, hence $q_{11} = 0, q_{12} = 1$ in w-coordinates. By (7.2.6)

$$A_{12} = 0,$$

and w are local conformal principal curvature coordinates. □

Conclusion of the proof of Theorem 7.2.1:
By the above proposition, we see in the degenerate case that $\dim \mathcal{V}_f = \dim \mathcal{T} - 1$, and we select $e \perp \mathcal{V}_f, \neq 0$ such that $T_{\pi g}\mathcal{T} = \mathcal{V}_f \oplus \text{span}\{e\}$. In local conformal principal curvature coordinates, as established in the above proposition, we construct two variations $V_\pm \in C^\infty(\Sigma, \mathbb{R}^n)$ such that for the second variation

$$\pm\langle \delta^2 \pi_f . V_\pm, e \rangle > 0.$$

Firther we select variations $V_1, \ldots, V_{\dim \mathcal{T}-1}$ with

$$\text{span}\{\delta\pi_f . V_1, \ldots, \delta\pi_f . V_{\dim \mathcal{T}-1}\} = \mathcal{V}_f.$$

Then by implicit function theorem for t small, there exists $\lambda \in \mathbb{R}^{\dim \mathcal{T}-1}$ such that

$$\pi\Big((f + tV + \lambda_r V_r)^* g_{\text{euc}}\Big) = \pi(f^* g_{\text{euc}}) + \alpha e.$$

For $\pm\alpha \geq 0$, there exists $\mu_\pm \in \mathbb{R}, \tilde{\lambda} \sim \lambda$ such that

$$\pi\Big((f + tV + \tilde{\lambda}_r V_r + \mu_\pm V_\pm)^* g_{\text{euc}}\Big) = \pi(f^* g_{\text{euc}}),$$

and the whole-filling procedure applies to prove smoothness and the Euler-Lagrange equation (7.1.1) for f as in the full rank case. This concludes the proof of the theorem. □

Note added after the lecture:
Recently, the first author and Li and independently Riviere have extended the existence of constrained minimizers for $\mathcal{W}(\Sigma, g_0, n) < 8\pi$ in any codimension, see [KuLi10] and [Ri10]. Whereas smoothness of the minimizer and the Euler-Lagrange equation (7.1.1) is obtained in all cases along the lines presented in the above lecture, this is not addressed in [KuLi10], and is left open in [Ri10] in the degenerate case. Even more recent the second author has extended the framework of this section in [Sch11] to $\mathcal{W}(\Sigma, g_0, n) < 8\pi$ in any codimension.

Appendix

7.A The space S_2^{TT}

Definition 7.A.1. Let Σ be an open surface and g a metric on Σ. The space $S_2^{TT}(g)$ of transverse traceless tensors is defined to be the set of smooth 2-covariant tensors which are symmetric, traceless, transverse that is

$$\begin{aligned} q_{ij} &= q_{ji}, \\ \operatorname{tr}_g q &= g^{ij} q_{ij} = 0, \\ \operatorname{div}_g q_k &= g^{ij} \nabla_i q_{jk} = 0. \end{aligned}$$

□

For conformal metrics S_2^{TT} is the same.

Proposition 7.A.1. *For $\tilde{g} = e^{2u} g$ we have $S_2^{TT}(\tilde{g}) = S_2^{TT}(g)$.*

Proof. Clearly a 2-covariant tensor is symmetric or tracefree with respect to $\tilde{g}$ if and olny if it is symmetric or tracefree with respect to g.

We recall from Proposition 1.2.2 for the difference of the Christoffel symbols which is always a tensor that

$$T_{ij}^k := \tilde{\Gamma}_{ij}^k - \Gamma_{ij}^k = \delta_i^k \partial_j u + \delta_j^k \partial_i u - g_{ij} g^{kl} \partial_l u$$

in any coordinates. We continue

$$\begin{aligned} \operatorname{div}_{\tilde{g}} q_k &= \tilde{g}^{ij} \nabla_i^{\tilde{g}} q_{jk} = \tilde{g}^{ij} (\partial_i q_{jk} - \tilde{\Gamma}_{ij}^m q_{mk} - \tilde{\Gamma}_{ik}^m q_{jm}) \\ &= \tilde{g}^{ij} \nabla_i^g q_{jk} - \tilde{g}^{ij} (T_{ij}^m q_{mk} + T_{ik}^m q_{jm}) \\ &= e^{-2u} \operatorname{div}_g q_k - e^{-2u} g^{ij} (T_{ij}^m q_{mk} + T_{ik}^l q_{jm}). \end{aligned}$$

When we prove

$$g^{ij} (T_{ij}^m q_{mk} + T_{ik}^m q_{jm}) = 0 \tag{7.A.1}$$

for all symmetric and tracefree q, we get

$$\operatorname{div}_{\tilde{g}} q_k = 0 \Longleftrightarrow \operatorname{div}_g q = 0,$$

and the proposition follows.

Now by the formula above

$$\begin{aligned} g^{ij}(T_{ij}^m q_{mk} + T_{ik}^m q_{jm}) &= g^{ij}(\delta_i^m \partial_j u + \delta_j^m \partial_i u - g_{ij} g^{ml} \partial_l u) q_{mk} \\ &\quad + g^{ij}(\delta_i^m \partial_k u + \delta_k^m \partial_i u - g_{ik} g^{ml} \partial_l u) q_{jm} \\ &= g^{ij} q_{ik} \partial_j u + g^{ij} q_{jk} \partial_i u - 2 g^{ml} q_{mk} \partial_l u \\ &\quad + g^{ij} q_{ij} \partial_k u + g^{ij} q_{jk} \partial_i u - g^{ml} q_{km} \partial_l u \\ &= 0, \end{aligned}$$

which is (7.A.1) and the proposition is proved. □

Proposition 7.A.2. *Let q be a 2-covariant symmetric, tracefree tensor with respect to a metric g on an open surface Σ. Then q is divergence free if and only if in local conformal coordinates the function*

$$q_{11} - i q_{12} \text{ is holomorphic.}$$

Proof. Since in conformal coordinates $g = e^{2u} g_{\text{euc}}$ by the previous proposition $S_2^{TT}(g) = S_2^{TT}(g_{\text{euc}})$, we have to prove

$$q \in S_2^{TT}(g_{\text{euc}}) \Longleftrightarrow q_{11} - i q_{12} \text{ is holomorphic}$$

for q symmetric and tracefree with respect to g_{euc} that is $q_{12} = q_{21}, q_{11} = -q_{22}$. Indeed

$$\begin{aligned} & q \in S_2^{TT}(g_{\text{euc}}) \\ \Longleftrightarrow \quad & \delta^{ij} \partial_i q_{jk} = 0 \\ \Longleftrightarrow \quad & \begin{cases} \partial_1 q_{11} + \partial_2 q_{21} = 0 \\ \partial_1 q_{12} + \partial_2 q_{22} = 0 \end{cases} \\ \Longleftrightarrow \quad & \begin{cases} \partial_1 q_{11} + \partial_2 q_{12} = 0 \\ \partial_2 q_{11} - \partial_1 q_{12} = 0 \end{cases} \end{aligned}$$

which are exactly the Cauchy Riemann equations for $q_{11} - i q_{12}$, hence are equivalent to $q_{11} - i q_{12}$ being holomorphic. □

References

[KuLi10] E. KUWERT and X. LI, $W^{2,2}$-conformal immersions of a closed Riemann surface into $\mathbb{R}^n$, 2010, arXiv:math.DG/1007.3967.

[KuSch07] E. KUWERT and R. SCHÄTZLE, Minimizers of the Willmore functional under fixed conformal class, 2007, arXiv:math.DG/1009.6168.

[KuSch11] E. KUWERT and R. SCHÄTZLE, *Closed surfaces with bounds on their Willmore energy*, to appear in Annali della Scuola Normale Superiore di Pisa, 2011 arXiv:math.DG/1009.5286.

[LY82] P. LI and S. T. YAU, *A new conformal invariant and its applications to the Willmore conjecture and the first eigenvalue on compact surfaces*, Inventiones Mathematicae, **69** (1982), 269–291.

[MoRo86] S. MONTIEL and A. ROS, *Minimal immersions of surfaces by the first Eigenfunctions and conformal Area*, Inventiones Mathematicae **83** (1986), 153–166.

[Tr] A. TROMBA, "Teichmüller Theory in Riemannian Geometry", Birkhäuser, 1992.

[Ri10] T. RIVIÈRE, *Variational Principles for immersed Surfaces with L^2-bounded Second Fundamental Form*, 2010, arXiv:math.AP/1007.2997.

[Sch11] R. SCHÄTZLE, *Estimation of the conformal factor under bounded Willmore energy*, (2011), preprint.

8 The large genus limit of the infimum of the Willmore energy

8.1 The large genus limit

In this talk we are concerned with the behaviour of the minimal Willmore energy

$$\beta_p^n := \inf\{W(f) \mid f : \Sigma \to \mathbb{R}^n, \text{genus}(\Sigma) = p, \Sigma \text{ orientable } \},$$

for closed, orientable surfaces of fixed genus, as defined in remark 1. after Theorem 6.1.1 when the genus tends to infinity. We have seen in Proposition 2.1.1 that $\mathcal{W}(f) \geq 4\pi$ and equality only for round spheres. Firstly this gives $\beta_0^n = 4\pi$. Secondly, the existence of a minimizer for closed, orientable surfaces of fixed genus in $\mathbb{R}^n$ was proved in [Sim93] when combined with [BaKu03], hence $\beta_p^n > 4\pi$ for $p \geq 1$. On the other hand the constrcution in remark 2. after Theorem 6.1.1 shows $\beta_p^n \leq 8\pi$. This can be improved to strict inequality. Indeed by (1.3.12) the Willmore energy of a minimal surface in S^n equals its area, and equals by conformal invariance in Proposition 1.2.3 the Willmore energy of its stereographic projection in $\mathbb{R}^n$. Now, Kühnel and Pinkall [KuPi86] and independently Kusner [Kus87,Kus89] observed that the minimal surfaces $\xi_{p,1} \subseteq S^3$ of genus p described by Lawson in [La70] have area less than 8π. In summary we know that

$$4\pi < \beta_p^n < 8\pi \quad \text{for } p \geq 1.$$

The aim of this talk is to determine the limit of β_p^n when the genus tends to infinity.

Theorem 8.1.1 ([KuLiSch09]).

$$\lim_{p\to\infty} \beta_p^n = 8\pi.$$

Proof. By the estimate above, we know already that

$$\limsup_{p\to\infty} \beta_p^n \leq 8\pi,$$

and we have to prove that

$$\beta_\infty^n := \liminf_{p\to\infty} \beta_p^n \geq 8\pi.$$

Suppose not, that is $\beta_\infty^n < 8\pi$. We select $p_k \to \infty$ with $\beta_{p_k}^n \to \beta_\infty^n$ and by the existence of a minimizer for closed, orientable surfaces of fixed genus let $\Sigma_k \subseteq \mathbb{R}^n$ be an embedded, closed, orientable Willmore surfaces with

$$\begin{aligned} \text{genus}(\Sigma_k) &= p_k, \\ \mathcal{W}(\Sigma_k) &= \beta^n_{p_k}. \end{aligned}$$

We can restrict to embedded surfaces instead of immersions by Proposition 1.2.3, as $\beta^n_{p_k} < 8\pi$ for large k. Translating and rescaling Σ_k, we may assume $0 \in \Sigma_k$ with

$$\| A_{\Sigma_k} \|_{L^\infty(\Sigma_k)} \leq |A_{\Sigma_k}(0)| = 1.$$

By the area bound in (2.1.15), we see $\varrho^{-2}\mathcal{H}^2(\Sigma_k \cap B_\varrho) \leq C\mathcal{W}(\Sigma_k) \leq C$ and get for ϱ small enough that $\int_{\Sigma_k \cap B_{2\varrho}} |A_{\Sigma_k}| \, d\mathcal{H}^2 \leq \varepsilon_0$, hence by interior estimates in Theorem 3.3.1 that

$$\| \nabla^{\perp,k} A_{\Sigma_k} \|_{L^\infty(\Sigma_k)} \leq C_k \varrho^{-k-1}.$$

Then for a subsequence $id_{\Sigma_k} \to f_\infty$ smoothly after reparametrization on compact subsets of $\mathbb{R}^n$ and $\mathcal{H}^2 \llcorner \Sigma_k \to \mu_{f_\infty}$ weakly as Radon measures. Then by monotonicity formula (2.1.24)

$$\begin{aligned} \#(f^{\infty,-1}(.)) &\leq \theta^2(\mu_{f_\infty}, \cdot) \leq \theta^2(\mu_{f_\infty}, \infty) + \frac{1}{4\pi}\mathcal{W}(f_\infty) \\ &\leq \liminf_{k \to \infty} \frac{1}{4\pi}\mathcal{W}(\Sigma_k) = \liminf_{k \to \infty} \beta^n_{p_k}/(4\pi) \qquad (8.1.1) \\ &= \beta^n_\infty/(4\pi) < 2. \end{aligned}$$

We see that f_∞ is an embedding, $\mu_{f_\infty} = \mathcal{H}^2 \llcorner \Sigma_\infty$ and

$$\begin{gathered} \Sigma_k \to \Sigma_\infty \quad \text{smoothly on compact subsets of } \mathbb{R}^n, \\ \Sigma_\infty \text{ is a smooth, oriented, embedded Willmore surface,} \qquad (8.1.2) \\ \| A_{\Sigma_\infty} \|_{L^\infty(\Sigma_\infty)} \leq |A_{\Sigma_\infty}(0)| = 1. \end{gathered}$$

As $\text{genus}(\Sigma_k) = p_k \to \infty$ and by Proposition 2.1.4, we get that

$$\Sigma_\infty \text{ is non-compact and connected.}$$

To continue, we need the following Lemma which is obtained by Feder's dimension reduction argument.

Lemma 8.1.1. *Let* $M \subseteq \mathbb{R}^n$ *be an oriented, minimal surface with* $\partial M = 0$ *and*

$$\theta^2(\mathcal{H}^2 \llcorner M, \infty) < 2.$$

Then $M = 0$ *or* M *is a unit density plane.* □

Firstly, we conclude that Σ_∞ is not minimal, that is

$$\mathcal{W}(\Sigma_\infty) > 0, \tag{8.1.3}$$

since otherwise by the lemma above and (8.1.1), we concldue that $A_{\Sigma_\infty} \equiv 0$, which contradicts $A_{\Sigma_\infty}(0) \neq 0$ of (8.1.2). Secondly, we need the following proposition.

Proposition 8.1.1. *For* $r_j \to \infty$ *we get after passing to a subsequence*

$$r_j^{-1}\Sigma_\infty \to P \quad \text{smoothly in compact subsets of } \mathbb{R}^n - \{0\},$$

where P *is a plane containing the origin which possibly depends on the subsequence.*

Proof. The smooth convergence is the main step to be proved. Clearly, P is oriented, and $P \cap B_\varrho(0) - B_{\varrho/2}(0) \neq 0$ for all $\varrho > 0$, since Σ_∞ is connected and non-compact, hence $P \neq 0$ and $0 \in P$.

We see for $\varrho > 0$

$$\begin{aligned}\mathcal{W}(P - B_\varrho(0)) &\leq \limsup_{r\to\infty} \mathcal{W}(r^{-1}\Sigma_\infty - B_\varrho(0)) \\ &= \limsup_{r\to\infty} \mathcal{W}(\Sigma_\infty - B_{r\varrho}(0)) = 0,\end{aligned}$$

as $\mathcal{W}(\Sigma_\infty) < \infty$, hence $\mathcal{W}(P) = 0$ and P is minimal. Moreover by monotonicity formula (2.1.24), we get recalling (8.1.1)

$$\theta^2(\mathcal{H}^2 \llcorner P, \infty) \leq \limsup_{r\to\infty} \left(\theta^2(\mathcal{H}^2 \llcorner r^{-1}\Sigma_\infty, \infty) + \frac{1}{4\pi}\mathcal{W}(r^{-1}\Sigma_\infty)\right)$$

$$= \theta^2(\mathcal{H}^2 \llcorner \Sigma_\infty, \infty) + \frac{1}{4\pi}\mathcal{W}(\Sigma_\infty) < 2.$$

Then by the previous lemma, P is a unit density plane. □

Now we replace Σ_∞ by $r_j^{-1}\Sigma_\infty$ and may assume by (8.1.3)

$$\mathcal{W}(\Sigma_\infty \cap B_{1/2}(0)) \geq \mathcal{W}(\Sigma_\infty) - \varepsilon$$

for j large and ε small. By the previous Proposition, we see that Σ_∞ is smoothly close to a unit density plane in $B_2(0) - B_{1/2}(0)$ for j large.

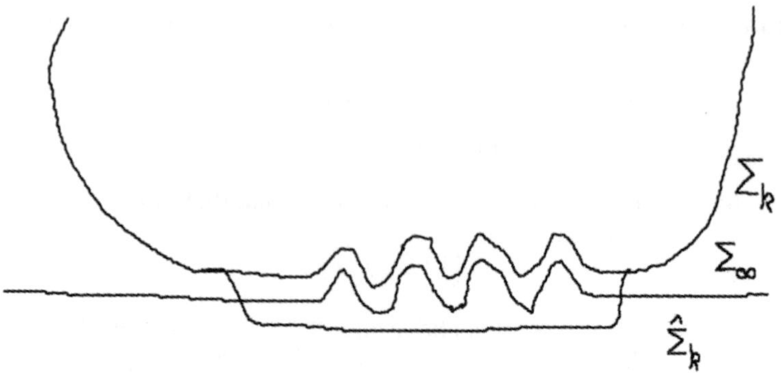

Since $\Sigma_k \to \Sigma_\infty$, we know that

$$\Sigma_k \text{ is smoothly close to a unit density plane in } B_2(0) - B_{1/2}(0),$$

$$\mathcal{W}(\Sigma_k \cap B_1(0)) \geq \mathcal{W}(\Sigma_\infty \cap B_1(0)) - \varepsilon \geq \mathcal{W}(\Sigma_\infty) - 2\varepsilon,$$

$$\text{genus}(\Sigma_k \cap B_1(0)) = \text{genus}(\Sigma_\infty \cap B_1(0)) \leq \text{genus}(\Sigma_\infty) < \infty,$$

for k large, where we write with slight abuse of notation $\text{genus}(\Sigma_k \cap B_1(0)) = \text{genus}((\Sigma_k \cap B_1(0)) \oplus B_1(0))$.

Therefore, we can replace Σ_k in $B_1(0)$ by a surface close to a disc and obtain a closed, orientable surface $\hat{\Sigma}_k$ with

$$\hat{\Sigma}_k - B_1(0) = \Sigma_k - B_1(0),$$

$$\mathcal{W}(\hat{\Sigma}_k \cap B_1(0)) \leq \varepsilon \ll \mathcal{W}(\Sigma_\infty) - 2\varepsilon \leq \mathcal{W}(\Sigma_k \cap B_1(0)),$$

$$\text{genus}(\hat{\Sigma}_k) = \text{genus}(\Sigma_k) - \text{genus}(\Sigma_k \cap B_1(0)) \geq p_k - \text{genus}(\Sigma_\infty).$$

This yields $\text{genus}(\hat{\Sigma}_k) \to \infty$, hence

$$\begin{aligned} \beta^n_\infty &= \liminf_{p\to\infty} \beta^n_p \leq \liminf_{k\to\infty} \mathcal{W}(\hat{\Sigma}_k) \\ &\leq \liminf_{k\to\infty} \mathcal{W}(\Sigma_k) - (\mathcal{W}(\Sigma_\infty) - 3\varepsilon) \\ &= \beta^n_\infty - (\mathcal{W}(\Sigma_\infty) - 3\varepsilon) < \beta^n_\infty, \end{aligned}$$

by (8.1.3) if $3\varepsilon < \mathcal{W}(\Sigma_\infty)$, which is a contradiction, and hence proves the theorem. □

References

[BaKu03] M. Bauer and E. Kuwert, *Existence of Minimizing Willmore Surfaces of Prescribed Genus*, IMRN Intern. Math. Res. Notes, **10** (2003), 553–576.

[KuPi86] W. KÜHNEL and U. PINKALL, *On total mean curvature*, Quarterly of Applied Mathematics, **37** (1986), 437–447.

[Kus87] R. KUSNER, Global geometry of extremal surfaces in three-space, Dissertation University of California, Berkeley, 1987.

[Kus89] R. KUSNER, *Comparison Surfaces for the Willmore problem*, Pacific Journal of Mathematics, **138** (1989), 317–345.

[KuLiSch09] E. KUWERT, X. LI and R. SCHÄTZLE, The large genus limit of the infimum of the Willmore energy, American Journal of Mathematics (2009), to appear.

[KuSch01] E. KUWERT and R. SCHÄTZLE, *The Willmore Flow with small initial energy*, Journal of Differential Geometry, **57** (2001), 409–441.

[La70] H.B. LAWSON, *Complete minimal surfaces in S^3*, Annals of Mathematics, **92** (1970), 335–374.

[LY82] P. LI and S.T. YAU, *A new conformal invariant and its applications to the Willmore conjecture and the first eigenvalue on compact surfaces*, Inventiones Mathematicae **69** (1982), 269–291.

[Sim93] L. SIMON, *Existence of surfaces minimizing the Willmore functional*, Communications in Analysis and Geometry, **1** (1993), 281–326.

The role of conservation laws in the analysis of conformally invariant problems

Tristan Rivière

Contents **117**

1 Introduction

These lecture notes form the cornerstone between two areas of Mathematics: calculus of variations and conformal invariance theory.

Conformal invariance plays a significant role in many areas of Physics, such as conformal field theory, renormalization theory, turbulence, general relativity. Naturally, it also plays an important role in geometry: theory of Riemannian surfaces, Weyl tensors, Q-curvature, Yang-Mills fields, etc... We shall be concerned with the study of conformal invariance in *analysis*. More precisely, we will focus on the study of nonlinear

PDEs arising from conformally invariant variational problems (*e.g.* harmonic maps, prescribed mean curvature surfaces, Yang-Mills equations, amongst others).

A transformation is called conformal when it preserves angles infinitesimally, that is, when its differential is a similarity at every point. Unlike in higher dimensions, the group of conformal transformations in two dimensions is very large; it has infinite dimension. In fact, it contains as many elements as there are holomorphic and antiholomorphic maps. This particularly rich feature motivates us to restrict our attention on the two-dimensional case. Although we shall not be concerned with higher dimension, the reader should know that many of the results presented in these notes can be generalized to any dimension.

The first historical instance in which calculus of variations encountered conformal invariance took place early in the twentieth century with the resolution of the Plateau problem. Originally posed by J.-L. Lagrange in 1760, it was solved independently over 150 years later by J. Douglas and T. Radó. In recognition of his work, the former was bestowed the first Fields Medal in 1936 (jointly with L. Alhfors).

Plateau Problem. Given a regular closed connected curve Γ in $\mathbb{R}^3$, does there exist an immersion u of the unit-disk D^2 such that ∂D^2 is homeomorphically sent onto Γ and for which $u(D^2)$ has a minimal area?

One of the most important ideas introduced by Douglas and Radó consists in minimizing the energy of the map u

$$E(u) = \frac{1}{2}\int_{D^2} |\partial_x u|^2 + |\partial_y u|^2 \, dx \wedge dy.$$

It has good coercivity properties and lower semicontinuity in the weak topology of the Sobolev space $W^{1,2}(D^2, \mathbb{R}^3)$, unlike the area functional

$$A(u) = \int_{D^2} |\partial_x u \times \partial_y u| \, dx \wedge dy.$$

One crucial observation is the following inequality, valid for all u in $W^{1,2}(D^2, \mathbb{R}^3)$,

$$A(u) \le E(u),$$

with equality if and only if u is weakly conformal, namely:

$$|\partial_x u| = |\partial_y u| \quad \text{and} \quad \partial_x u \cdot \partial_y u = 0 \quad \text{a.e.}$$

The energy functional E has another advantage over the area functional A. While A is invariant under the action of the *infinite* group of diffeomorphisms of D^2 into itself[1], the functional E is only invariant through the action of the much smaller group (it is in fact *finite*) of Möbius transformations comprising conformal, degree 1 maps from D^2 into itself[2].

In effect, the idea of Douglas and Radó bears resemblance to that of minimizing, in a normal parametrization $|\dot{\gamma}| = 1$, the energy of a curve $\int_{[0,1]} |\dot{\gamma}|^2\, dt$, rather than the Lagrangian of the arclength $\int_{[0,1]} |\dot{\gamma}|\, dt$, which is invariant under the too big group of positive diffeomorphisms of the segment $[0, 1]$.

All the disks (D^2, g) are conformally equivalent to the flat disk D^2. Thus, the aforementioned observations enable us to infer that any minimum of the area functional A, if it exists, must be a critical point of the energy functional E. These points are the harmonic maps u in $\mathbb{R}^3$ satisfying

$$\Delta u = 0 \qquad \text{in} \quad \mathcal{D}'(D^2). \tag{1.1}$$

Leading this process to fruition is however hindered by the boundary data, which is of a "free" Dirichlet type along a curve Γ, and by the noncompacity of the Möbius group, which will thus have to be "broken" by

[1] Indeed, given two distinct positive parametrizations (x, y) and (x', y') of the unit-disk D^2, there holds, for each pair of functions f and g on D^2, the identity

$$df \wedge dg = \partial_x f \partial_y g - \partial_y f \partial_x g\, dx \wedge dy = \partial_{x'} f \partial_{y'} g - \partial_{y'} f \partial_{x'} g\, dx' \wedge dy'$$

so that, owing to $dx \wedge dy$ and $dx' \wedge dy'$ having the same sign, we find

$$|\partial_x f \partial_y g - \partial_y f \partial_x g|\, dx \wedge dy = |\partial_{x'} f \partial_{y'} g - \partial_{y'} f \partial_{x'} g|\, dx' \wedge dy'.$$

This implies that A is invariant through composition with positive diffeomorphisms.

[2] The invariance of E under conformal transformations may easily be seen by working with the complex variable $z = x + iy$. Indeed, we note

$$\partial_z := \frac{1}{2}\left(\partial_x - i\partial_y\right)$$

and

$$\partial_{\overline{z}} := \frac{1}{2}\left(\partial_x + i\partial_y\right)$$

so that $du = \partial_z u\, dz + \partial_{\overline{z}} u\, d\overline{z}$, and thus

$$E(u) = \frac{i}{2}\int_{D^2} |\partial_z u|^2 + |\partial_{\overline{z}} u|^2\, dz \wedge d\overline{z}.$$

Accordingly, if we compose u with a conformal transformation, *i.e.* holomorphic, $z = f(w)$, there holds for $\tilde{u}(w) = u(z)$ the identities

$$|\partial_w \tilde{u}|^2 = |f'(w)|^2\, |\partial_z u|^2 \circ f \quad \text{and} \quad |\partial_{\overline{w}} \tilde{u}|^2 = |f'(w)|^2\, |\partial_{\overline{z}} u|^2 \circ f.$$

Moreover, $dz \wedge d\overline{z} = |f'(w)|^2\, dw \wedge d\overline{w}$. Bringing altogether these results yields the desired conformal invariance $E(u) = E(\tilde{u})$.

the so-called *three-point method.* Eventually, one reaches the following result.

Theorem 1.1 (Douglas-Radó-Courant). *Given a regular closed curve Γ in $\mathbb{R}^3$, there exists a continuous minimum u for the energy E within the space of $W^{1,2}(D^2,\mathbb{R}^3)$ functions mapping the boundary of the unit-disk ∂D^2 onto Γ in a monotone fashion, and satisfying*

$$\begin{cases} \Delta u = 0 & \text{in} \quad D^2 \\ |\partial_x u|^2 - |\partial_y u|^2 - 2i\, \partial_x u \cdot \partial_y u = 0 & \text{in} \quad D^2. \end{cases} \tag{1.2}$$

The harmonicity and conformality condition exhibited in (1.2) implies that $u(D^2)$ realizes a minimal surface[3]. R. Osserman showed that it has no branch points in the interior of the unit-disk. This result was subsequently generalized to the boundary of the disk by S. Hildebrandt.

The resolution of the Plateau problem proposed by Douglas and Radó is an example of the use of a conformal invariant Lagrangian E to approach an "extrinsic" problem: minimizing the area of a disk with fixed boundary. The analysis of this problem was eased by the high simplicity of the equation (1.1) satisfied by the critical points of E. It is the Laplace equation. Hence, questions related to unicity, regularity, compactness, etc... can be handled with a direct application of the maximum principle. In these lecture notes, we will be concerned with analogous problems (in particular regularity issues) related to the critical points of conformally invariant, coercive Lagrangians with quadratic growth. As we will discover, the maximum principle no longer holds, and one must seek an alternate way to compensate this lack. The conformal invariance of the Lagrangian will generate a very peculiar type of nonlinearities in the corresponding Euler-Lagrange equations. We will see how the specific structure of these nonlinearities enables one to recast the equations in divergence form. This new formulation, combined to the results of *integration by compensation*, will provide the substrate to understanding a variety of problems, such as Willmore surfaces, poly-harmonic and

[3] Recall the following result from differential geometry. Let u be a positive conformal parametrization from an oriented disk in $\mathbb{R}^3$. The mean curvature vector $\vec{H}$, parallel to the outward unit normal vector $\vec{n}$, is defined as

$$\vec{H} = H\,\vec{n} = 2^{-1}\,e^{-2\lambda}\,\Delta u,$$

where $e^{\lambda} = |\partial_x u| = |\partial_y u|$ and H is the mean curvature $H = (\kappa_1 + \kappa_2)/2$. Equivalently, there holds

$$\Delta u = 2H\,\partial_x u \times \partial_y u. \tag{1.3}$$

α-harmonic maps, Yang-Mills fields, Hermite-Einstein equations, wave maps, etc...

ACKNOWLEDGEMENTS. These notes have been written at the occasion of several mini-courses given by the author in the spring and summer 2009 successively at the mathematics Department of Nice, of Warwick University, at the PIMS-Vancouver and at the De Giorgi Institute in Pisa. He would like to thank these institutes for the excellent working conditions they provided during these teaching periods which have been very stimulating and fruitful experiences for him.

2 Conformally invariant coercive Lagrangians with quadratic growth, in dimension 2

We consider a Lagrangian of the form

$$L(u) = \int_{D^2} l(u, \nabla u)\, dx\, dy, \tag{2.1}$$

where the integrand l is a function of the variables $z \in \mathbb{R}^m$ and $p \in \mathbb{R}^2 \otimes \mathbb{R}^m$, which satisfies the following coercivity and "almost quadratic" conditions in p:

$$C^{-1}\, |p|^2 \leq l(z, p) \leq C\, |p|^2. \tag{2.2}$$

We further assume that L is conformally invariant: for each positive conformal transformation f of degree 1, and for each map $u \in W^{1,2}(D^2,\mathbb{R}^m)$, there holds

$$\begin{aligned} L(u \circ f) &= \int_{f^{-1}(D^2)} l(u \circ f, \nabla(u \circ f))\, dx'\, dy' \\ &= \int_{D^2} l(u, \nabla u)\, dx\, dy = L(u). \end{aligned} \tag{2.3}$$

Example 2.1. The Dirichlet energy described in the Introduction,

$$E(u) = \int_{D^2} |\nabla u|^2\, dx\, dy,$$

whose critical points satisfy the Laplace equation (1.1), owing to the conformal hypothesis, geometrically describes *minimal surfaces*. Regularity and compactness matters relative to this equation are handled with the help of the maximum principle.

Example 2.2. Let an arbitrary metric in $\mathbb{R}^m$ be given, namely $(g_{ij})_{i,j\in\mathbb{N}_m}\in C^1(\mathbb{R}^m, \mathcal{S}_m^+)$, where $\mathcal{S}_m^+$ denotes the subset of $M_m(\mathbb{R})$, comprising the symmetric positive definite $m \times m$ matrices. We make the following uniform coercivity and boundedness hypothesis:

$$\exists\ C > 0 \qquad \text{such that} \quad C^{-1}\delta_{ij} \leq g_{ij} \leq C\delta_{ij} \quad \text{on } \mathbb{R}^m.$$

Finally, we suppose that

$$\|\nabla g\|_{L^\infty(\mathbb{R}^m)} < +\infty.$$

With these conditions, the second example of quadratic, coercive, conformally invariant Lagrangian is

$$\begin{aligned} E_g(u) &= \frac{1}{2}\int_{D^2} \langle \nabla u, \nabla u\rangle_g \, dx\, dy \\ &= \frac{1}{2}\int_{D^2} \sum_{i,j=1}^m g_{ij}(u)\nabla u^i \cdot \nabla u^j \, dx\, dy. \end{aligned}$$

Note that Example 2.1 is contained as a particular case.

Verifying that E_g is indeed conformally invariant may be done analogously to the case of the Dirichlet energy, via introducing the complex variable $z = x + iy$. No new difficulty arises, and the details are left to the reader as an exercise.

The weak critical points of E_g are the functions $u \in W^{1,2}(D^2, \mathbb{R}^m)$ which satisfy

$$\forall \xi \in C_0^\infty(D^2, \mathbb{R}^m) \qquad \frac{d}{dt} E_g(u + t\xi)_{|_{t=0}} = 0.$$

An elementary computation reveals that u is a weak critical point of E_g if and only if the following Euler-Lagrange equation holds in the sense of distributions:

$$\forall i = 1 \cdots m \qquad \Delta u^i + \sum_{k,l=1}^m \Gamma^i_{kl}(u)\, \nabla u^k \cdot \nabla u^l = 0. \tag{2.4}$$

Here, Γ^i_{kl} are the Christoffel symbols corresponding to the metric g, explicitly given by

$$\Gamma^i_{kl}(z) = \frac{1}{2}\sum_{s=1}^m g^{is}\left(\partial_{z_l} g_{ks} + \partial_{z_k} g_{ls} - \partial_{z_s} g_{kl}\right), \tag{2.5}$$

where (g^{ij}) is the inverse matrix of (g_{ij}).

Equation (2.4) bears the name *harmonic map equation*[4] with values in $(\mathbb{R}^m, g)$.

Just as in the flat setting, if we further suppose that u is conformal, then (2.4) is in fact equivalent to $u(D^2)$ being a minimal surface in $(\mathbb{R}^m, g)$.

We note that $\Gamma^i(\nabla u, \nabla u) := \sum_{k,l=1}^m \Gamma^i_{kl} \nabla u^k \cdot \nabla u^l$, so that the harmonic map equation can be recast as

$$\Delta u + \Gamma(\nabla u, \nabla u) = 0. \tag{2.6}$$

This equation raises several analytical questions:

(i) **Weak limits:** Let u_n be a sequence of solutions of (2.6) with uniformly bounded energy E_g. Can one extract a subsequence converging weakly in $W^{1,2}$ to a harmonic map?

(ii) **Palais-Smale sequences:** Let u_n be a sequence of solutions of (2.6) in $W^{1,2}(D^2, \mathbb{R}^m)$ with uniformly bounded energy E_g, and such that

$$\Delta u_n + \Gamma(\nabla u_n, \nabla u_n) = \delta_n \to 0 \qquad \text{strongly in } H^{-1}.$$

Can one extract a subsequence converging weakly in $W^{1,2}$ to a harmonic map?

(iii) **Regularity of weak solutions:** Let u be a map in $W^{1,2}(D^2, \mathbb{R}^m)$ which satisfies (2.4) distributionally. How regular is u? Continuous, smooth, analytic, etc...

The answer to (iii) is strongly tied to that of (i) and (ii). We shall thus restrict our attention in these notes on regularity matters.

Prior to bringing into light further examples of conformally invariant Lagrangians, we feel worthwhile to investigate deeper the difficulties associated with the study of the regularity of harmonic maps in two dimensions.

The harmonic map equation (2.6) belongs to the class of elliptic systems with *quadratic growth*, also known as *natural growth*, of the form

$$\Delta u = f(u, \nabla u), \tag{2.7}$$

where $f(z, p)$ is an arbitrary continuous function for which there exists constants $C_0 > 0$ and $C_1 > 0$ satisfying

$$\forall z \in \mathbb{R}^m \quad \forall p \in \mathbb{R}^2 \otimes \mathbb{R}^m \qquad f(z, p) \le C_1 |p|^2 + C_0. \tag{2.8}$$

---

[4] One way to interpret (2.4) as the two-dimensional equivalent of the geodesic equation in normal parametrization,

$$\frac{d^2 x^i}{dt^2} + \sum_{k,l=1}^m \Gamma^i_{kl} \frac{dx^k}{dt} \frac{dx^l}{dt} = 0.$$

In dimension two, these equations are critical for the Sobolev space $W^{1,2}$. Indeed,

$$u \in W^{1,2} \Rightarrow \Gamma(\nabla u, \nabla u) \in L^1 \Rightarrow \nabla u \in L^p_{\rm loc}(D^2) \quad \forall p < 2.$$

In other words, from the regularity standpoint, the demand that ∇u be square-integrable provides the information that[5] ∇u belongs to $L^p_{\rm loc}$ for all $p < 2$. We have thus lost a little bit of information ! Had this not been the case, the problem would be "boostrapable", thereby enabling a successful study of the regularity of u. Therefore, in this class of problems, the main difficulty lies in the aforementioned slight loss of information, which we symbolically represent by $L^2 \to L^{2,\infty}$.

There are simple examples of equations with quadratic growth in two dimensions for which the answers to the questions (i)-(iii) are all negative. Consider[6]

$$\Delta u + |\nabla u|^2 = 0. \tag{2.10}$$

This equation has quadratic growth, and it admits a solution in $W^{1,2}(D^2)$ which is unbounded in L^∞, and thus discontinuous. It is explicitly given by

$$u(x, y) := \log\log \frac{2}{\sqrt{x^2 + y^2}}.$$

The regularity issue can thus be answered negatively. Similarly, for the equation (2.10), it takes little effort to devise counter-examples to the weak limit question (i), and thus to the question (ii). To this end, it is helpful to observe that C^2 maps obey the general identity

$$\Delta e^u = e^u \left[\Delta u + |\nabla u|^2\right]. \tag{2.11}$$

---

[5] Actually, one can show that ∇u belongs to the weak-L^2 Marcinkiewicz space $L^{2,\infty}_{\rm loc}$ comprising those measurable functions f for which

$$\sup_{\lambda>0} \lambda^2 \left|\{p \in D^2 \ |f(p)| > \lambda\}\right| < +\infty, \tag{2.9}$$

where $|\cdot|$ is the standard Lebesgue measure. Note that $L^{2,\infty}$ is a slightly larger space than L^2. However, it possesses the same scaling properties.

[6] This equation is conformally invariant. However, as shown by J. Frehse [11], it is also the Euler-Lagrange equation derived from a Lagrangian which is *not* conformally invariant:

$$L(u) = \int_{D^2} \left(1 + \frac{1}{1 + e^{12u}\,(\log 1/|(x, y)|)^{-12}}\right) |\nabla u|^2(x, y)\, dx\, dy.$$

One easily verifies that if v is a positive solution of

$$\Delta v = -2\pi \sum_i \lambda_i\, \delta_{a_i},$$

where $\lambda_i > 0$ and δ_{a_i} are isolated Dirac masses, then $u := \log v$ provides a solution[7] in $W^{1,2}$ of (2.10). We then select a positive regular function f with integral equal to 1, and supported on the ball of radius $1/4$ centered on the origin. There exists a sequence of atomic measures with positive weights λ_i^n such that

$$f_n = \sum_{i=1}^{n} \lambda_i^n\, \delta_{a_i^n} \qquad \text{and} \qquad \sum_{i=1}^{n} \lambda_i^n = 1, \tag{2.12}$$

which converges as Radon measures to f. We next introduce

$$u_n(x, y) := \log\left[\sum_{i=1}^{n} \lambda_i^n \log \frac{2}{|(x, y) - a_i^n|}\right].$$

On D^2, we find that

$$v_n = \sum_{i=1}^{n} \lambda_i^n \log \frac{2}{|(x, y) - a_i^n|} > \sum_{i=1}^{n} \lambda_i^n \log \frac{8}{5} = \log \frac{8}{5}. \tag{2.13}$$

On the other hand, there holds

$$\begin{aligned} \int_{D^2} |\nabla u_n|^2 &= -\int_{D^2} \Delta u_n = -\int_{\partial D^2} \frac{\partial u_n}{\partial r} \\ &\leq \int_{\partial D^2} \frac{|\nabla v_n|}{|v_n|} \leq \frac{1}{\log \frac{8}{5}} \int_{\partial D^2} |\nabla v_n| \leq C \end{aligned}$$

[7] Indeed, per (2.11), we find $\Delta u + |\nabla u|^2 = 0$ away from the points a_i. Near these points, ∇u asymptotically behaves as follows:

$$|\nabla u| = |v|^{-1}\, |\nabla v| \simeq \left(|(x, y) - a_i|\, \log |(x, y) - a_i|\right)^{-1} \in L^2.$$

Hence, $|\nabla u|^2 \in L^1$, so that $\Delta u + |\nabla u|^2$ is a distribution in $H^{-1} + L^1$ supported on the isolated points a_i. ¿From this, it follows easily that

$$\Delta u + |\nabla u|^2 = \sum_i \mu_i\, \delta_{a_i}.$$

Thus, Δu is the sum of an L^1 function and of Dirac masses. But because Δu lies in H^{-1}, the coefficients μ_i must be zero. Accordingly, u does belong to $W^{1,2}$.

for some constant C independent of n. Hence, (u_n) is a sequence of solutions to (2.10) uniformly bounded in $W^{1,2}$. Since the sequence (f_n) converges as Radon measures to f, it follows that for any $p < 2$, the sequence (v_n) converges strongly in $W^{1,p}$ to

$$v := \log \frac{2}{r} * f.$$

The uniform lower bound (2.13) paired to the aforementioned strong convergence shows that for each $p < 2$, the sequence $u_n = \log v_n$ converges strongly in $W^{1,p}$ to

$$u := \log \left[\log \frac{2}{r} * f \right].$$

From the hypotheses satisfied by f, we see that $\Delta(e^u) = -2\pi\, f \neq 0$. As f is regular, so is thus e^u, and therefore, owing to (2.11), u cannot fullfill (2.10).

Accordingly, we have constructed a sequence of solutions to (2.10) which converges weakly in $W^{1,2}$ to a map that is *not* a solution to (2.10).

Example 2.3. We consider a map $(\omega_{ij})_{i,j\in\mathbb{N}_m}$ in $C^1(\mathbb{R}^m, so(m))$, where $so(m)$ is the space antisymmetric square $m \times m$ matrices. We impose the following uniform bound

$$\|\nabla\omega\|_{L^\infty(D^2)} < +\infty.$$

For maps $u \in W^{1,2}(D^2, \mathbb{R}^m)$, we introduce the Lagrangian

$$E^\omega(u) = \frac{1}{2}\int_{D^2} |\nabla u|^2 + \sum_{i,j=1}^m \omega_{ij}(u)\partial_x u^i \partial_y u^j - \partial_y u^i \partial_x u^j \, dx\, dy. \quad (2.14)$$

The conformal invariance of this Lagrangian arises from the fact that E^ω is made of the conformally invariant Lagrangian E to which is added the integral over D^2 of the 2-form $\omega = \omega_{ij} dz^i \wedge dz^j$ *pulled back* by u. Composing u by an arbitrary positive diffeomorphism of D^2 will not affect this integral, thereby making E^ω into a conformally invariant Lagrangian.

The Euler-Lagrange equation deriving from (2.14) for variations of the form $u+t\xi$, where ξ is an arbitrary smooth function with compact support in D^2, is found to be

$$\Delta u^i - 2\sum_{k,l=1}^m H^i_{kl}(u)\, \nabla^\perp u^k \cdot \nabla u^l = 0 \qquad \forall\ i = 1\cdots m. \quad (2.15)$$

Here, $\nabla^\perp u^k = (-\partial_y u^k, \partial_x u^k)$[8] while H^i_{kl} is antisymmetric in the indices k and l. It is the coefficient of the $\mathbb{R}^m$-valued two-form H on $\mathbb{R}^m$

$$H^i(z) := \sum_{k,l=1}^{m} H^i_{kl}(z)\, dz^k \wedge dz^l.$$

The form H appearing in the Euler-Lagrange equation (2.15) is the unique solution of

$$\begin{aligned} \forall z \in \mathbb{R}^m \quad \forall U, V, W \in \mathbb{R}^m \\ d\omega_z(U, V, W) &= 4\, U \cdot H(V, W) \\ &= 4 \sum_{i=1}^{m} U^i\, H^i(V, W). \end{aligned}$$

For instance, in dimension three, $d\omega$ is a 3-form which can be identified with a function on $\mathbb{R}^m$. More precisely, there exists H such that $d\omega = 4H\, dz^1 \wedge dz^2 \wedge dz^3$. In this notation (2.15) may be recast, for each $i \in \{1, 2, 3\}$, as

$$\Delta u^i = 2H(u)\, \partial_x u^{i+1} \partial_y u^{i-1} - \partial_x u^{i-1} \partial_y u^{i+1}, \tag{2.16}$$

where the indexing is understood in $\mathbb{Z}_3$. The equation (2.16) may also be written

$$\Delta u = 2H(u)\, \partial_x u \times \partial_y u,$$

which we recognize as the *prescribed mean curvature equation.*

In a general fashion, the equation (2.15) admits the following geometric interpretation. Let u be a conformal solution of (2.15), so that $u(D^2)$ is a surface whose mean curvature vector at the point (x, y) is given by

$$e^{-2\lambda}\, u^* H = \left(e^{-2\lambda} \sum_{k,l=1}^{m} H^i_{kl}(u)\, \nabla^\perp u^k \cdot \nabla u^l \right)_{i=1\cdots m}, \tag{2.17}$$

where e^λ is the conformal factor $e^\lambda = |\partial_x u| = |\partial_y u|$. As in Example 2.2, the equation (2.15) forms an elliptic system with quadratic growth, thus critical in dimension two for the $W^{1,2}$ norm. The analytical difficulties relative to this nonlinear system are thus, *a priori*, of the same nature as those arising from the *harmonic map equation.*

[8] In our notation, $\nabla^\perp u^k \cdot \nabla u^l$ is the Jacobian

$$\nabla^\perp u^k \cdot \nabla u^l = \partial_x u^k \partial_y u^l - \partial_y u^k \partial_x u^l.$$

Example 2.4. In this last example, we combine the settings of Examples 2.2 and 2.3 to produce a mixed problem. Given on $\mathbb{R}^m$ a metric g and a two-form ω, both C^1 with uniformly bounded Lipschitz norm, consider the Lagrangian

$$E_g^\omega(u) = \frac{1}{2}\int_{D^2} \langle \nabla u, \nabla u\rangle_g \; dx\, dy + u^*\omega.$$

As before, it is a coercive conformally invariant Lagrangian with quadratic growth. Its critical points satisfy the Euler-Lagrange equation

$$\Delta u^i + \sum_{k,l=1}^m \Gamma^i_{kl}(u)\nabla u^k \cdot \nabla u^l - 2\sum_{k,l=1}^m H^i_{kl}(u)\nabla^\perp u^k \cdot \nabla u^l = 0, \quad (2.18)$$

for $i = 1 \cdots m$.

Once again, this elliptic system admits a geometric interpretation which generalizes the ones from Examples 2.2 and 2.3. Whenever a conformal map u satisfies (2.18), then $u(D^2)$ is a surface in $(\mathbb{R}^m, g)$ whose mean curvature vector is given by (2.17). The equation (2.18) also forms an elliptic system with quadratic growth, and critical in dimension two for the $W^{1,2}$ norm.

Interestingly enough, M. Grüter showed that *any* coercive conformally invariant Lagrangian with quadratic growth is of the form E_g^ω for some appropriately chosen g and ω.

Theorem 2.5 ([15]). *Let $l(z, p)$ be a real-valued function on $\mathbb{R}^m \times \mathbb{R}^2 \otimes \mathbb{R}^m$, which is C^1 in its first variable and C^2 in its second variable. Suppose that l obeys the coercivity and quadratic growth conditions*

$$\begin{aligned} &\exists C > 0 \quad \text{such that} \quad \forall z \in \mathbb{R}^m \quad \forall p \in \mathbb{R}^2 \otimes \mathbb{R}^m \\ &\qquad C^{-1}|p|^2 \le l(z, p) \le C|p|^2. \end{aligned} \quad (2.19)$$

Let L be the Lagrangian

$$L(u) = \int_{D^2} l(u, \nabla u)(x, y)\, dx\, dy \quad (2.20)$$

acting on $W^{1,2}(D^2, \mathbb{R}^m)$-maps u. We suppose that L is conformally invariant: for every conformal application ϕ positive and of degree 1, *there holds*

$$L(u \circ \phi) = \int_{\phi^{-1}(D^2)} l(u \circ \phi, \nabla(u \circ \phi))(x, y)\, dx\, dy = L(u). \quad (2.21)$$

Then there exist on $\mathbb{R}^m$ a C^1 metric g and a C^1 two-form ω such that

$$L = E_g^\omega. \quad (2.22)$$

Maps taking values in a submanifold of $\mathbb{R}^m$

Up to now, we have restricted our attention to maps from D^2 into a manifold with only one chart $(\mathbb{R}^n, g)$. More generally, it is possible to introduce the Sobolev space $W^{1,2}(D^2, N^n)$, where (N^n, g) is an oriented n-dimensional C^2-manifold. When this manifold is compact without boundary (which we shall henceforth assume, for the sake of simplicity), a theorem by Nash guarantees that it can be isometrically immersed into Euclidean space $\mathbb{R}^m$, for m large enough. We then define

$$W^{1,2}(D^2, N^n) := \left\{u \in W^{1,2}(D^2, \mathbb{R}^m) \;\; u(p) \in N^n \;\text{ a.e. } p \in D^2\right\}.$$

Given on N^n a C^1 two-form ω, we may consider the Lagrangian

$$E^\omega(u) = \frac{1}{2}\int_{D^2} |\nabla u|^2 \, dx\, dy + u^*\omega \tag{2.23}$$

acting on maps $u \in W^{1,2}(D^2, N^n)$. The critical points of E^ω are defined as follows. Let π_N be the orthogonal projection on N^n which to each point in a neighborhood of N associates its nearest orthogonal projection on N^n. For points sufficiently close to N, the map π_N is regular. We decree that $u \in W^{1,2}(D^2, N^n)$ is a critical point of E^ω whenever there holds

$$\frac{d}{dt} E^\omega(\pi_N(u + t\xi))_{t=0} = 0, \tag{2.24}$$

for all $\xi \in C_0^\infty(D^2, \mathbb{R}^m)$.

It can be shown[9] that (2.24) is satisfied by $u \in C_0^\infty(D^2, \mathbb{R}^m)$ if and only if u obeys the Euler-Lagrange equation

$$\Delta u + A(u)(\nabla u, \nabla u) = H(u)(\nabla^\perp u, \nabla u), \tag{2.25}$$

where A $(\equiv A_z)$ is the second fundamental form at the point $z \in N^n$ corresponding to the immersion of N^n into $\mathbb{R}^m$. To a pair of vectors in $T_z N^n$, the map A_z associates a vector orthogonal to $T_z N^n$. In particular, at a point $(x, y) \in D^2$, the quantity $A_{(x,y)}(u)(\nabla u, \nabla u)$ is the vector of $\mathbb{R}^m$ given by

$$A_{(x,y)}(u)(\nabla u, \nabla u) := A_{(x,y)}(u)(\partial_x u, \partial_x u) + A_{(x,y)}(u)(\partial_y u, \partial_y u).$$

For notational convenience, we henceforth omit the subscript (x, y).

[9] In codimension 1, this is done below.

Similarly, $H(u)(\nabla^\perp u, \nabla u)$ at the point $(x, y) \in D^2$ is the vector in $\mathbb{R}^m$ given by

$$H(u)(\nabla^\perp u, \nabla u) := H(u)(\partial_x u, \partial_y u) - H(u)(\partial_y u, \partial_x u)$$
$$= 2H(u)(\partial_x u, \partial_y u),$$

where H ($\equiv H_z$) is the $T_z N^n$-valued alternating two-form on $T_z N^n$:

$$\forall\, U, V, W \in T_z N^n \qquad d\omega(U, V, W) := U \cdot H_z(V, W).$$

Note that in the special case when $\omega = 0$, the equation (2.25) reduces to

$$\Delta u + A(u)(\nabla u, \nabla u) = 0, \tag{2.26}$$

which is known as the *N^n-valued harmonic map equation.*

We now establish (2.25) in the codimension 1 case. Let ν be the normal unit vector to N. The form ω may be naturally extended on a small neighborhood of N^n via the pull-back $\pi_N^*\omega$ of the projection π_N. Infinitesimally, to first order, considering variations for E^ω of the form $\pi_N(u+t\xi)$ is tantamount to considering variations of the kind $u + t\, d\pi_N(u)\xi$, which further amounts to focusing on variations of the form $u + tv$, where $v \in W^{1,2}(D^2, \mathbb{R}^m) \cap L^\infty$ satisfies $v \cdot \nu(u) = 0$ almost everywhere. Following the argument from Example 2.3, we obtain that u is a critical point of E^ω whenever for all v with $v \cdot \nu(u) = 0$ a.e., there holds

$$\int_{D^2} \sum_{i=1}^{m} \left[\Delta u^i - 2 \sum_{k,l=1}^{m} H^i_{kl}(u)\; \nabla^\perp u^k \cdot \nabla u^l \right] v^i \, dx\, dy = 0,$$

where H is the vector-valued two-form on $\mathbb{R}^m$ given for z on N^n by

$$\forall\, U, V, W \in \mathbb{R}^m \qquad d\pi_N^*\omega(U, V, W) := U \cdot H_z(V, W).$$

In the sense of distributions, we thus find that

$$\left[\Delta u - H(u)(\nabla^\perp u, \nabla u)\right] \wedge \nu(u) = 0. \tag{2.27}$$

Recall, $\nu \circ u \in L^\infty \cap W^{1,2}(D^2, \mathbb{R}^m)$. Accordingly (2.27) does indeed make sense in $\mathcal{D}'(D^2)$.

Note that if any of the vectors U, V and W is normal to N^n, *i.e.* parallel to ν, then $d\pi_N^*\omega(U, V, W) = 0$, so that

$$\nu_z \cdot H_z(V, W) = 0 \qquad \forall\, V, W \in \mathbb{R}^m.$$

Whence,

$$\begin{aligned}\left[\Delta u - H(u)(\nabla^{\perp}u, \nabla u)\right] \cdot \nu(u) &= \Delta u \cdot \nu(u) \\ &= \operatorname{div}(\nabla u \cdot \nu(u)) - \nabla u \cdot \nabla(\nu(u)) \\ &= -\nabla u \cdot \nabla(\nu(u))\end{aligned} \quad (2.28)$$

where we have used the fact that $\nabla u \cdot \nu(u) = 0$ holds almost everywhere, since ∇u is tangent to N^n.

Altogether, (2.27) and (2.28) show that u satisfies in the sense of distributions the equation

$$\Delta u - H(u)(\nabla^{\perp}u, \nabla u) = -\nu(u)\ \nabla(\nu(u)) \cdot \nabla u. \quad (2.29)$$

In codimension 1, the second fundamental form acts on a pair of vectors (U, V) in $T_z N^n$ via

$$A_z(U, V) = \nu(z)\ \ < d\nu_z U, V >, \quad (2.30)$$

so that, as announced, (2.29) and (2.25) are identical.

We close this section by stating a conjecture formulated by Stefan d'Hildebrandt in the late 1970s.

Conjecture 2.6 ([20,21]). The critical points with finite energy of a coercive conformally invariant Lagrangian with quadratic growth are Hölder continuous.

The remainder of these lecture notes shall be devoted to establishing this conjecture. Although its resolution is closely related to the compactness questions (i) and (ii) previously formulated on page 9, for lack of time, we shall not dive into the study of this point.

Our proof will begin by recalling the first partial answers to Hildebrandt's conjecture provided by H. Wente and F. Hélein, and the importance in their approach of the rôle played by *conservations laws* and *integration by compensation*.

Then, in the last section, we will investigate the theory of linear elliptic systems with antisymmetric potentials, and show how to apply it to the resolution of Hildebrandt's conjecture.

3 Integrability by compensation applied to the regularity of critical points of some conformally invariant Lagrangians

3.1 Constant mean curvature equation (CMC)

Let $H \in \mathbb{R}$ be constant. We study the analytical properties of solutions in $W^{1,2}(D^2, \mathbb{R}^3)$ of the equation

$$\Delta u - 2H\ \partial_x u \times \partial_y u = 0. \quad (3.1)$$

The Jacobian structure of the right-hand side enables without much trouble, inter alia, to show that **Palais-Smale sequences** converge weakly:

Let F_n be a sequence of distributions converging to zero in $H^{-1}(D^2,\mathbb{R}^3)$, and let u_n be a sequence of functions uniformly bounded in $W^{1,2}$ and satisfying the equation

$$\Delta u_n - 2H\ \partial_x u_n \times \partial_y u_n = F_n \to 0 \text{ strongly in } H^{-1}(D^2).$$

We use the notation

$$\begin{aligned}(\partial_x u_n \times \partial_y u_n)^i &= \partial_x u_n^{i+1} \partial_y u_n^{i-1} - \partial_x u_n^{i-1} \partial_y u_n^{i+1} \\ &= \partial_x (u_n^{i+1}\ \partial_y u_n^{i-1}) - \partial_y (u_n^{i+1} \partial_x u_n^{i-1}).\end{aligned} \tag{3.2}$$

The uniform bounded on the $W^{1,2}$-norm of u_n enables the extraction of a subsequence $u_{n'}$ weakly converging in $W^{1,2}$ to some limit u_∞. With the help of the Rellich-Kondrachov theorem, we see that the sequence u_n is strongly compact in L^2. In particular, we can pass to the limit in the following quadratic terms

$$u_n^{i+1}\ \partial_y u_n^{i-1} \to u_\infty^{i+1}\ \partial_y u_\infty^{i-1} \qquad \text{in } \mathcal{D}'(D^2)$$

and

$$u_n^{i+1}\ \partial_x u_n^{i-1} \to u_\infty^{i+1}\ \partial_x u_\infty^{i-1} \qquad \text{in } \mathcal{D}'(D^2).$$

Combining this to (3.2) reveals that u_∞ is a solution of the CMC equation (3.1).

Obtaining information on the regularity of weak $W^{1,2}$ solutions of the CMC equation (3.2) requires some more elaborate work. More precisely, a result from the theory of integration by compensation due to H. Wente is needed.

Theorem 3.1 ([36]). *Let a and b be two functions in $W^{1,2}(D^2)$, and let ϕ be the unique solution in $W_0^{1,p}(D^2)$ – for $1 \le p < 2$ – of the equation*

$$\begin{cases} -\Delta\phi = \partial_x a\, \partial_y b - \partial_x b\, \partial_y a & \textit{in } D^2 \\ \varphi = 0 & \textit{on } \partial D^2. \end{cases} \tag{3.3}$$

Then ϕ belongs to $C^0 \cap W^{1,2}(D^2)$ and

$$\|\phi\|_{L^\infty(D^2)} + \|\nabla\phi\|_{L^2(D^2)} \le C_0\ \|\nabla a\|_{L^2(D^2)}\ \|\nabla b\|_{L^2(D^2)}, \tag{3.4}$$

where C_0 is a constant independent of a and b.[10]

[10] Actually, one shows that Theorem 3.1 may be generalized to arbitrary oriented Riemannian surfaces, with a constant C_0 *independent of the surface*, which is quite a remarkable and useful fact. For more details, see [13] and [34].

Proof of Theorem 3.1. We shall first assume that a and b are smooth, so as to legitimize the various manipulations which we will need to perform. The conclusion of the theorem for general a and b in $W^{1,2}$ may then be reached through a simple density argument. In this fashion, we will obtain the continuity of ϕ from its being the uniform limit of smooth functions.

Observe first that integration by parts and a simple application of the Cauchy-Schwarz inequality yields the estimate

$$\begin{aligned}\int_{D^2} |\nabla\phi|^2 = -\int_{D^2} \phi\,\Delta\phi &\leq \|\phi\|_\infty\ \|\partial_x a\,\partial_y b - \partial_x b\,\partial_y a\|_1 \\ &\leq 2\,\|\phi\|_\infty \|\nabla a\|_2\ \|\nabla b\|_2.\end{aligned}$$

Accordingly, if ϕ lies in L^∞, then it automatically lies in $W^{1,2}$.

Given two functions $\tilde{a}$ and $\tilde{b}$ in $C_0^\infty(\mathbb{C})$, which is dense in $W^{1,2}(\mathbb{C})$, we first establish the estimate (3.4) for

$$\tilde{\phi} := \frac{1}{2\pi}\log\frac{1}{r} * \left[\partial_x\tilde{a}\,\partial_y\tilde{b} - \partial_x\tilde{b}\,\partial_y\tilde{a}\right]. \tag{3.5}$$

Owing to the translation-invariance, it suffices to show that

$$|\tilde{\phi}(0)| \leq C_0\ \|\nabla\tilde{a}\|_{L^2(\mathbb{C})}\ \|\nabla\tilde{b}\|_{L^2(\mathbb{C})}. \tag{3.6}$$

We have

$$\begin{aligned}\tilde{\phi}(0) &= -\frac{1}{2\pi}\int_{\mathbb{R}^2} \log r\ \partial_x\tilde{a}\,\partial_y\tilde{b} - \partial_x\tilde{b}\,\partial_y\tilde{a} \\ &= -\frac{1}{2\pi}\int_0^{2\pi}\int_0^{+\infty} \log r\ \frac{\partial}{\partial r}\left(\tilde{a}\,\frac{\partial\tilde{b}}{\partial\theta}\right) - \frac{\partial}{\partial\theta}\left(\tilde{a}\,\frac{\partial\tilde{b}}{\partial r}\right)\,dr\,d\theta \\ &= \frac{1}{2\pi}\int_0^{2\pi}\int_0^{+\infty} \tilde{a}\,\frac{\partial\tilde{b}}{\partial\theta}\,\frac{dr}{r}\,d\theta.\end{aligned}$$

Since $\int_0^{2\pi}\frac{\partial\tilde{b}}{\partial\theta}\,d\theta = 0$, we may deduct from each circle $\partial B_r(0)$ a constant $\tilde{a}$ chosen to have average $\bar{\tilde{a}}_r$ on $\partial B_r(0)$. Hence, there holds

$$\tilde{\phi}(0) = \frac{1}{2\pi}\int_0^{2\pi}\int_0^{+\infty} [\tilde{a} - \bar{\tilde{a}}_r]\,\frac{\partial\tilde{b}}{\partial\theta}\,\frac{dr}{r}\,d\theta.$$

Applying successively the Cauchy-Schwarz and Poincaré inequalities on the circle S^1, we obtain

$$|\tilde{\phi}(0)| \leq \frac{1}{2\pi}\int_0^{+\infty}\frac{dr}{r}\left(\int_0^{2\pi}|\tilde{a}-\bar{\tilde{a}}_r|^2\right)^{\frac{1}{2}}\left(\int_0^{2\pi}\left|\frac{\partial\tilde{b}}{\partial\theta}\right|^2\right)^{\frac{1}{2}}$$
$$\leq \frac{1}{2\pi}\int_0^{+\infty}\frac{dr}{r}\left(\int_0^{2\pi}\left|\frac{\partial\tilde{a}}{\partial\theta}\right|^2\right)^{\frac{1}{2}}\left(\int_0^{2\pi}\left|\frac{\partial\tilde{b}}{\partial\theta}\right|^2\right)^{\frac{1}{2}}.$$

The sought after inequality (3.6) may then be inferred from the latter via applying once more the Cauchy-Schwarz inequality.

Returning to the disk D^2, the Whitney extension theorem yields the existence of $\tilde{a}$ and $\tilde{b}$ such that

$$\int_{\mathbb{C}}|\nabla\tilde{a}|^2 \leq C_1\int_{D^2}|\nabla a|^2 \tag{3.7}$$

and

$$\int_{\mathbb{C}}|\nabla\tilde{b}|^2 \leq C_1\int_{D^2}|\nabla b|^2. \tag{3.8}$$

Let $\tilde{\phi}$ be the function in (3.5). The difference $\phi-\tilde{\phi}$ satisfies the equation

$$\begin{cases}\Delta(\phi-\tilde{\phi})=0 & \text{in } D^2\\ \phi-\tilde{\phi}=-\tilde{\phi} & \text{on } \partial D^2.\end{cases}$$

The *maximum principle* applied to the inequalities (3.6), (3.7) and (3.8) produces

$$\|\phi-\tilde{\phi}\|_{L^\infty(D^2)} \leq \|\tilde{\phi}\|_{L^\infty(\partial D^2)} \leq C\|\nabla a\|_2\,\|\nabla b\|_2.$$

With the triangle inequality $|\|\phi\|_\infty - \|\tilde{\phi}\|_\infty| \leq \|\phi-\tilde{\phi}\|_\infty$ and the inequality (3.6), we reach the desired L^∞-estimate of ϕ, and therefore, per the above discussion, the theorem is proved. □

Proof of the regularity of the solutions of the CMC equation

Our first aim will be to establish the existence of a positive constant α such that

$$\sup_{\rho<1/4,\ p\in B_{1/2}(0)} \rho^{-\alpha}\int_{B_\rho(p)}|\nabla u|^2 < +\infty. \tag{3.9}$$

Owing to a classical result from Functional Analysis[11], the latter implies that $u \in C^{0,\alpha/2}(B_{1/2}(0))$. From this, we deduce that u is locally Hölder continuous in the interior of the disk D^2. We will then explain how to obtain the smoothness of u from its Hölder continuity.

Let $\varepsilon_0 > 0$. There exists some radius $\rho_0 > 0$ such that for every $r < \rho_0$ and every point p in $B_{1/2}(0)$

$$\int_{B_r(p)} |\nabla u|^2 < \varepsilon_0.$$

We shall in due time adjust the value ε_0 to fit our purposes. In the sequel, $r < \rho_0$. On $B_r(p)$, we decompose $u = \phi + v$ in such a way that

$$\begin{cases} \Delta\phi = 2H\ \partial_x u \times \partial_y u & \text{in} \quad B_r(p) \\ \phi = 0 & \text{on} \quad \partial B_r(p). \end{cases}$$

Applying Theorem 3.1 to ϕ yields

$$\begin{aligned} \int_{B_r(p)} |\nabla\phi|^2 &\leq C_0|H| \int_{B_r(p)} |\nabla u|^2 \int_{B_r(p)} |\nabla u|^2 \\ &\leq C_0|H|\ \varepsilon_0 \int_{B_r(p)} |\nabla u|^2. \end{aligned} \tag{3.10}$$

The function $v = u - \phi$ is harmonic. To obtain useful estimates on v, we need the following result.

Lemma 3.2. *Let v be a harmonic function on D^2. For every point p in D^2, the function*

$$\rho \longmapsto \frac{1}{\rho^2} \int_{B_\rho(p)} |\nabla v|^2$$

is increasing.

Proof. Note first that

$$\frac{d}{d\rho}\left[\frac{1}{\rho^2}\int_{B_\rho(p)} |\nabla v|^2\right] = -\frac{2}{\rho^3}\int_{B_\rho(p)} |\nabla v|^2 + \frac{1}{\rho^2}\int_{\partial B_\rho(p)} |\nabla v|^2. \tag{3.11}$$

Denote by $\overline{v}$ the average of v on $\partial B_\rho(p)$: $\overline{v} := |\partial B_\rho(p)|^{-1} \int_{\partial B_\rho(p)} v$. Then, there holds

$$0 = \int_{B_\rho(p)} (v - \overline{v})\ \Delta v = -\int_{B_\rho(p)} |\nabla v|^2 + \int_{\partial B_\rho(p)} (v - \overline{v})\ \frac{\partial v}{\partial \rho}.$$

[11] See for instance [14].

This implies that

$$\frac{1}{\rho}\int_{B_\rho(p)} |\nabla v|^2 \le \left(\frac{1}{\rho^2}\int_{\partial B_\rho(p)} |v-\overline{v}|^2\right)^{\frac{1}{2}} \left(\int_{\partial B_\rho(p)} \left|\frac{\partial v}{\partial \rho}\right|^2\right)^{\frac{1}{2}}. \tag{3.12}$$

In Fourier space, v satisfies $v = \sum_{n\in\mathbb{Z}} a_n\, e^{in\theta}$ and $v-\overline{v} = \sum_{n\in\mathbb{Z}^*} a_n\, e^{in\theta}$. Accordingly,

$$\frac{1}{2\pi\rho}\int_{\partial B_\rho(p)} |v-\overline{v}|^2 = \sum_{n\in\mathbb{Z}^*} |a_n|^2 \le \sum_{n\in\mathbb{Z}^*} |n|^2 |a_n|^2 \le \frac{1}{2\pi}\int_0^{2\pi} \left|\frac{\partial v}{\partial \theta}\right|^2 d\theta.$$

Combining the latter with (3.12) then gives

$$\frac{1}{\rho}\int_{B_\rho(p)} |\nabla v|^2 \le \left(\int_{\partial B_\rho(p)} \left|\frac{1}{\rho}\frac{\partial v}{\partial \theta}\right|^2\right)^{\frac{1}{2}} \left(\int_{\partial B_\rho(p)} \left|\frac{\partial v}{\partial \rho}\right|^2\right)^{\frac{1}{2}}. \tag{3.13}$$

If we multiply the Laplace equation throughout by $(x-x_p)\,\partial_x v + (y-y_p)\,\partial_y v$, and then integrate by parts over $B_\rho(p)$, we reach the **Pohozaev identity**:

$$2\int_{\partial B_\rho(p)} \left|\frac{\partial v}{\partial \rho}\right|^2 = \int_{\partial B_\rho(p)} |\nabla v|^2. \tag{3.14}$$

Altogether with (3.13), this identity implies that the right-hand side of (3.11) is positive, thereby concluding the proof [12]. □

We now return to the proof of the regularity of the solutions of the CMC equation. Per the above lemma, there holds

$$\int_{B_{\rho/2}(p)} |\nabla v|^2 \le \frac{1}{4}\int_{B_\rho(p)} |\nabla v|^2. \tag{3.15}$$

Since $\Delta v = 0$ on $B_\rho(p)$, while $\phi = 0$ on $\partial B_\rho(p)$, we have

$$\int_{B_\rho(p)} \nabla v \cdot \nabla\phi = 0.$$

[12] Another proof of Lemma 3.2 goes as follows: if v is harmonic then $f := |\nabla v|^2$ is sub-harmonic – $\Delta|\nabla v|^2 \ge 0$ – and an elementary calculation shows that for any non negative subharmonic function f in $\mathbb{R}^n$ one has $d/dr(r^{-n}\int_{B_r} f) \ge 0$.

Combining this identity to the inequality in (3.15), we obtain

$$\int_{B_{\rho/2}(p)} |\nabla(v+\phi)|^2 \leq \frac{1}{2}\int_{B_\rho(p)} |\nabla(v+\phi)|^2 + 3\int_{B_\rho(p)} |\nabla\phi|^2, \tag{3.16}$$

which, accounting for (3.10), yields

$$\int_{B_{\rho/2}(p)} |\nabla u|^2 \leq \left(\frac{1}{2} + 3\, C_0\, |H|\, \varepsilon_0\right)\int_{B_\rho(p)} |\nabla u|^2. \tag{3.17}$$

If we adjust ε_0 sufficiently small as to have $3\, C_0\, |H|\, \varepsilon_0 < 1/4$, it follows that

$$\int_{B_{\rho/2}(p)} |\nabla u|^2 \leq \frac{3}{4}\int_{B_\rho(p)} |\nabla u|^2. \tag{3.18}$$

Iterating this inequality gives the existence of a constant $\alpha > 0$ such that for all $p \in B_{1/2}(0)$ and all $r < \rho$, there holds

$$\int_{B_r(p)} |\nabla u|^2 \leq \left(\frac{r}{\rho_0}\right)^\alpha \int_{D^2} |\nabla u|^2,$$

which implies (3.9). Accordingly, the solution u of the CMC equation is Hölder continuous.

Next, we infer from (3.9) and (3.1) the bound

$$\sup_{\rho<1/2,\ p\in B_{1/2}(0)} \rho^{-\alpha}\int_{B_\rho(p)} |\Delta u| < +\infty. \tag{3.19}$$

A classical estimate on Riesz potentials gives

$$|\nabla u|(p) \leq C\frac{1}{|x|} * \chi_{B_{1/2}}\, |\Delta u| + C \qquad \forall\ p \in B_{1/4}(0),$$

where $\chi_{B_{1/2}}$ is the characteristic function of the ball $B_{1/2}(0)$. Together with injections proved by Adams in [1], the latter shows that $u \in W^{1,q}(B_{1/4}(0))$ for any $q > (2-\alpha)/(1-\alpha)$. Substituted back into (3.1), this fact implies that $\Delta u \in L^r$ for some $r > 1$. The equation then becomes subcritical, and a standard bootstrapping argument eventually yields that $u \in C^\infty$. This concludes the proof of the regularity of solutions of the CMC equation.

3.2 Harmonic maps with values in the sphere S^n

When the target manifold N^n has codimension 1, the *harmonic map equation* (2.26) becomes (*cf.* (2.30))

$$-\Delta u = \nu(u)\, \nabla(\nu(u)) \cdot \nabla u, \tag{3.20}$$

where u still denotes the normal unit-vector to the submanifold $N^n \subset \mathbb{R}^{n+1}$. In particular, if N^n is the sphere S^n, there holds $\nu(u) = u$, and the equation reads

$$-\Delta u = u\, |\nabla u|^2. \tag{3.21}$$

Another characterization of (3.21) states that the function $u \in W^{1,2}(D^2, S^n)$ satisfies (3.21) if and only if

$$u \wedge \Delta u = 0 \qquad \text{in } \mathcal{D}'(D^2). \tag{3.22}$$

Indeed, any S^n-valued map u obeys

$$0 = \Delta \frac{|u|^2}{2} = \operatorname{div}(u\, \nabla u) = |\nabla u|^2 + u\, \Delta u$$

so that Δu is parallel to u as in (3.22) if and only if the proportionality is $-|\nabla u|^2$. This is equivalent to (3.21). Interestingly enough, J. Shatah [28] observed that (3.22) is tantamount to

$$\forall i, j = 1 \cdots n+1 \qquad \operatorname{div}(u^i\, \nabla u_j - u_j\, \nabla u^i) = 0. \tag{3.23}$$

This formulation of the equation for S^n-valued harmonic maps enables one to pass to the weak limit, just as we previously did in the CMC equation.

The **regularity of S^n-valued harmonic maps** was obtained by F. Hélein, [19]. It is established as follows.

For each pair of indices (i, j) in $\{1 \cdots n+1\}^2$, the equation (3.23) reveals that the vector field $u^i\, \nabla u^j - u^j\, \nabla u^i$ forms a curl term, and hence there exists $B^i_j \in W^{1,2}$ with

$$\nabla^\perp B^i_j = u^i\, \nabla u_j - u_j\, \nabla u^i.$$

In local coordinates, (3.21) may be written

$$-\Delta u^i = \sum_{j=1}^{n+1} u^i\, \nabla u_j \cdot \nabla u^j. \tag{3.24}$$

We then make the field $\nabla^{\perp} B^i_j$ appear on the right-hand side by observing that

$$\sum_{j=1}^{n+1} u_j \nabla u^i \cdot \nabla u^j = \nabla u^i \cdot \nabla \left(\sum_{j=1}^{n+1} |u^j|^2/2 \right) = \nabla u^i \cdot \nabla |u|^2/2 = 0.$$

Deducting this null term from the right-hand side of (3.24) yields that for all $i = 1 \cdots n+1$, there holds

$$\begin{aligned} -\Delta u^i &= \sum_{j=1}^{n+1} \nabla^{\perp} B^i_j \cdot \nabla u^j \\ &= \sum_{j=1}^{n+1} \partial_x B^i_j \, \partial_y u^j - \partial_y B^i_j \, \partial_x u^j . \end{aligned} \tag{3.25}$$

We recognize the same Jacobian structure which we previously employed to establish the regularity of solutions of the CMC equation. It is thus possible to adapt mutatis mutandis our argument to (3.25) so as to infer that S^n-valued harmonic maps are regular.

3.3 Hélein's moving frames method and the regularity of harmonic maps mapping into a manifold

When the target manifold is no longer a sphere (or, more generally, when it is no longer homogeneous), the aforementioned Jacobian structure disappears, and the techniques we employed no longer seem to be directly applicable.

To palliate this lack of structure, and thus extend the regularity result to harmonic maps mapping into an arbitrary manifold, F. Hélein devised the *moving frames method.* The divergence-form structure being the result of the global symmetry of the target manifold, Hélein's idea consists in expressing the harmonic map equation in preferred moving frames, called *Coulomb frames*, thereby compensating for the lack of global symmetry with "infinitesimal symmetries".

This method, although seemingly unnatural and rather mysterious, has subsequently proved very efficient to answer regularity and compactness questions, such as in the study of nonlinear wave maps (see [12, 29–31]). For this reason, it is worthwhile to dwell a bit more on Hélein's method. We first recall the main result of F. Hélein.

Theorem 3.3 ([19]). *Let N^n be a closed C^2-submanifold of $\mathbb{R}^m$. Suppose that u is a harmonic map in $W^{1,2}(D^2, N^n)$ that weakly satisfies the harmonic map equation* (2.26). *Then u lies in $C^{1,\alpha}$ for all $\alpha < 1$.*

Proof of Theorem 3.3 when N^n is a two-torus

The notion of *harmonic coordinates* has been introduced first in general relativity by Yvonne Choquet-Bruaht in the early fifties. She discovered that the formulation of Einstein equation in these coordinates simplifies in a spectacular way. This idea of searching optimal charts among all possible "gauges" has also been very efficient for harmonic maps into manifolds. Since the different works of Hildebrandt, Karcher, Kaul, Jäger, Jost, Widman...etc. in the seventies it was known that the intrinsic harmonic map system (2.18) becomes for instance almost "triangular" in harmonic coordinates $(x^\alpha)_\alpha$ in the target which are minimizing the Dirichlet energy $\int_U |dx^\alpha|^2_g \, dvol_g$. The drawback of this approach is that working with harmonic coordinates requires to localize in the target and to restrict only to maps taking values into a *single chart* in which such coordinates exist! While looking at regularity question this assumption is very restrictive as long as we don't know that the harmonic map u is continuous for instance. It is not excluded *a priori* that the weak harmonic map u we are considering "covers the whole target" even locally in the domain.

The main idea of Frederic Hélein was to extend the notion of *harmonic coordinates* of Choquet-Bruhat by replacing it with the more flexible harmonic or *Coulomb orthonormal frame*, notion for which no localization in the target is needed anymore. Precisely the Coulomb orthonormal frames are mappings $e = (e_1 \cdots e_n)$ from the domain D^2 into the orthonormal basis of $T_u N^n$ minimizing the Dirichlet energy but for the covariant derivatives D_g in the target:

$$\int_{D^2} \sum_{i=1}^{n} |D_g e_i|^2 \, dx\, dy = \int_{D^2} \sum_{i,k=1}^{n} |(e_k, \nabla e_i)|^2 \, dx\, dy,$$

where $(\cdot, \cdot)$ denotes the canonical scalar product in $\mathbb{R}^m$.

We will consider the case when N^n is a two-dimensional parallelizable manifold (*i.e.* admitting a global basis of tangent vectors for the tangent space), namely a torus T^2 arbitrarily immersed into Euclidean space $\mathbb{R}^m$, for m large enough. The case of the two-torus is distinguished. Indeed, in general, if a harmonic map u takes its values in an immersed manifold N^n, then it is possible to lift u to a harmonic map $\tilde{u}$ taking values in a parallelizable torus $(S^1)^q$ of higher dimension. Accordingly, the argument which we present below can be analogously extended to a more general setting[13].

[13] Although the lifting procedure is rather technical. The details are presented in [19, Lemma 4.1.2].

Let $u \in W^{1,2}(D^2, T^2)$ satisfy weakly (2.26). We equip T^2 with a global, regular, positive orthonormal tangent frame field $(\varepsilon_1, \varepsilon_2)$. Let $\tilde{e} := (\tilde{e}_1, \tilde{e}_2) \in W^{1,2}(D^2, \mathbb{R}^m \times \mathbb{R}^m)$ be defined by the composition

$$\tilde{e}_i(x, y) := \varepsilon_i(u(x, y)).$$

The map $(\tilde{e})$ is defined on D^2 and it takes its values in the tangent frame field to T^2. Define the energy

$$\min_{\psi \in W^{1,2}(D^2,\mathbb{R})} \int_{D^2} |(e_1, \nabla e_2)|^2 \, dx\, dy, \tag{3.26}$$

where $(\cdot, \cdot)$ is the standard scalar product on $\mathbb{R}^m$, and

$$e_1(x, y) + ie_2(x, y) := e^{i\psi(x,y)} (\tilde{e}_1(x, y) + i\tilde{e}_2(x, y)).$$

We seek to optimize the map $(\tilde{e})$ by minimizing this energy over the $W^{1,2}(D^2)$-maps taking values in the space of rotations of the plane $\mathbb{R}^2 \simeq T_{u(x,y)}T^2$. Our goal is to seek a frame field as regular as possible in which the harmonic map equation will be recast. The variational problem (3.26) is well-posed, and it further admits a solution in $W^{1,2}$. Indeed, there holds

$$|(e_1, \nabla e_2)|^2 = |\nabla \psi + (\tilde{e}_1, \nabla \tilde{e}_2)|^2.$$

Hence, there exists a unique minimizer in $W^{1,2}$ which satisfies

$$0 = \operatorname{div} (\nabla \psi + (\tilde{e}_1, \nabla \tilde{e}_2)) = \operatorname{div}((e_1, \nabla e_2)). \tag{3.27}$$

A priori, $(e_1, \nabla e_2)$ belongs to L^2. But actually, thanks to the careful selection brought in by the variational problem (3.26), we shall discover that the frame field $(e_1, \nabla e_2)$ over D^2 lies in $W^{1,1}$, thereby improving the original L^2 belongingness[14]. Because the vector field $(e_1, \nabla e_2)$ is divergence-free, there exists some function $\phi \in W^{1,2}$ such that

$$(e_1, \nabla e_2) = \nabla^{\perp} \phi. \tag{3.29}$$

[14] Further yet, owing to a result of Luc Tartar [33], we know that $W^{1,1}(D^2)$ is continuously embedded in the Lorentz space $L^{2,1}(D^2)$, whose dual is the Marcinkiewicz weak-L^2 space $L^{2,\infty}(D^2)$, whose definition was recalled in (2.9). A measurable function f is an element of $L^{2,1}(D^2)$ whenever

$$\int_0^{+\infty} \left|\left\{p \in D^2 \ |f(p)| > \lambda\right\}\right|^{\frac{1}{2}} \, d\lambda < +\infty. \tag{3.28}$$

On the other hand, ϕ satisfies by definition

$$-\Delta\phi = (\nabla e_1, \nabla^{\perp} e_2) = \sum_{j=1}^{m} \partial_y e_1^j \partial_x e_2^j - \partial_x e_1^j \partial_y e_2^j. \tag{3.30}$$

The right-hand side of this elliptic equation comprises only Jacobians of elements of $W^{1,2}$. This configuration is identical to those previously encountered in our study of the constant mean curvature equation and of the equation of S^n-valued harmonic maps. In order to capitalize on this particular structure, we call upon an extension of Wente's Theorem 3.1 due to Coifman, Lions, Meyer, and Semmes.

Theorem 3.4 ([5]). *Let a and b be two functions in $W^{1,2}(D^2)$, and let ϕ be the unique solution in $W_0^{1,p}(D^2)$, for $1 \le p < 2$, of the equation*

$$\begin{cases} -\Delta\phi = \partial_x a \, \partial_y b - \partial_x b \, \partial_y a & \text{in } D^2 \\ \phi = 0 & \text{on } \partial D^2. \end{cases} \tag{3.31}$$

Then ϕ lies in $W^{2,1}$ and

$$\|\nabla^2\phi\|_{L^1(D^2)} \le C_1 \, \|\nabla a\|_{L^2(D^2)} \, \|\nabla b\|_{L^2(D^2)}, \tag{3.32}$$

where C_1 is a constant independent of a and b.[15]

Applying this result to the solution ϕ of (3.30) then reveals that $(e_1,\nabla e_2)$ is indeed an element of $W^{1,1}$.

We will express the harmonic map equation (2.26) in this particular Coulomb frame field, distinguished by its increased regularity. Note that (2.26) is equivalent to

$$\begin{cases} (\Delta u, e_1) = 0 \\ (\Delta u, e_2) = 0. \end{cases} \tag{3.33}$$

Using the fact that

$$\begin{aligned} &\partial_x u, \partial_y u \in T_u N^n = \text{vec}\{e_1, e_2\} \\ &(\nabla e_1, e_1) = (\nabla e_2, e_2) = 0 \\ &(\nabla e_1, e_2) + (e_1, \nabla e_2) = 0 \end{aligned}$$

[15] Theorem 3.1 is a corollary of Theorem 3.4 owing to the Sobolev embedding $W^{2,1}(D^2) \subset W^{1,2} \cap C^0$. In the same vein, Theorem 3.4 was preceded by two intermediary results. The first one, by Luc Tartar [32], states that the Fourier transform of $\nabla\phi$ lies in the Lorentz space $L^{2,1}$, which also implies Theorem 3.1. The second one, due to Stefan Müller, obtains the statement of Theorem 3.4 under the additional hypothesis that the Jacobian $\partial_x a \, \partial_y b - \partial_x b \, \partial_y a$ be positive.

we obtain that (3.33) may be recast in the form

$$\begin{cases} \mathrm{div}((e_1, \nabla u)) = -(\nabla e_2, e_1) \cdot (e_2, \nabla u) \\ \mathrm{div}((e_2, \nabla u)) = (\nabla e_2, e_1) \cdot (e_1, \nabla u). \end{cases} \tag{3.34}$$

On the other hand, there holds

$$\begin{cases} \mathrm{rot}((e_1, \nabla u)) = -(\nabla^\perp e_2, e_1) \cdot (e_2, \nabla u) \\ \mathrm{rot}((e_2, \nabla u)) = (\nabla^\perp e_2, e_1) \cdot (e_1, \nabla u). \end{cases} \tag{3.35}$$

We next proceed by introducing the Hodge decompositions in L^2 of the frames $(e_i, \nabla u)$, for $i \in \{1, 2\}$. In particular, there exist four functions C_i and D_i in $W^{1,2}$ such that

$$(e_i, \nabla u) = \nabla C_i + \nabla^\perp D_i.$$

Setting $W := (C_1, C_2, D_1, D_2)$, the identities (3.34) and (3.35) become

$$-\Delta W = \Omega \cdot \nabla W, \tag{3.36}$$

where Ω is the vector field valued in the space of 4×4 matrices defined by

$$\Omega = \begin{pmatrix} 0 & -\nabla^\perp \phi & 0 & -\nabla \phi \\ \nabla^\perp \phi & 0 & \nabla \phi & 0 \\ 0 & \nabla \phi & 0 & -\nabla^\perp \phi \\ -\nabla \phi & 0 & \nabla^\perp \phi & 0 \end{pmatrix}. \tag{3.37}$$

Since $\phi \in W^{2,1}$, the following Theorem 3.5 implies that ∇W, and hence ∇u, belong to L^p for some $p > 2$, thereby enabling the initialization of a bootstrapping argument analogous to that previously encountered in our study of the CMC equation. This procedure yields that u lies in $W^{2,q}$ for all $q < +\infty$. Owing to the standard Sobolev embedding theorem, it follows that $u \in C^{1,\alpha}$, which concludes the proof of the desired Theorem 3.3 in the case when the target manifold of the harmonic map u is the two-torus.

Theorem 3.5. *Let W be a solution in $W^{1,2}(D^2, \mathbb{R}^n)$ of the linear system*

$$-\Delta W = \Omega \cdot \nabla W, \tag{3.38}$$

where Ω is a $W^{1,1}$ vector field on D^2 taking values in the space of $n \times n$ matrices. Then W belongs to $W^{1,p}(B_{1/2}(0))$, for some $p > 2$. In particular, W is Hölder continuous[16] [17].

Proof of Theorem 3.5. Just as in the proof of the regularity of solutions of the CMC equation, we seek to obtain a Morrey-type estimate via the existence of some constant $\alpha > 0$ such that

$$\sup_{p\in B_{1/2}(0)\ ,\ 0<\rho<1/4} \rho^{-\alpha} \int_{B_\rho(p)} |\Delta W| < +\infty. \tag{3.39}$$

The statement of the theorem is then a corollary of an inequality involving Riesz potentials (*cf.* [1] and the CMC equation case on page137 above).

Let $\varepsilon_0 > 0$ be some constant whose size shall be in due time adjusted to fit our needs. There exists some radius ρ_0 such that for every $r < \rho_0$ and every point $p \in B_{1/2}(0)$, there holds

$$\|\Omega\|_{L^{2,1}(B_r(p))} < \varepsilon_0.$$

Note that we have used the aforementioned continuous injection $W^{1,1} \subset L^{2,1}$.

Henceforth, we consider $r < \rho_0$. On $B_r(p)$, we introduce the decomposition $W = \Phi + V$, with

$$\begin{cases} \Delta\Phi = \Omega \cdot \nabla W & \text{in } B_r(p) \\ \Phi = 0 & \text{on } \partial B_r(p). \end{cases}$$

A classical result on Riesz potentials (*cf.* [1]) grants the existence of a constant C_0 independent of r and such that

$$\begin{aligned} \|\nabla\Phi\|_{L^{2,\infty}(B_r(p))} &\leq C_0 \int_{B_r(p)} |\Omega \cdot \nabla W| \\ &\leq C_0 \|\Omega\|_{L^{2,1}(B_r(p))}\ \|\nabla W\|_{L^{2,\infty}(B_r(p))} \\ &\leq C_0\, \varepsilon_0\ \|\nabla W\|_{L^{2,\infty}(B_r(p))}. \end{aligned} \tag{3.40}$$

---

[16] The statement of Theorem 3.4 is optimal. To see this, consider $u = \log\log 1/r = W$. One verifies easily that $u \in W^{1,2}(D^2, T^2)$ satisfies weakly (2.26). Yet, $\Omega \equiv \nabla u$ fails to be $W^{1,1}$, owing to

$$\int_0^1 \frac{dr}{r \log\frac{1}{r}} = +\infty.$$

[17] The hypothesis $\Omega \in W^{1,1}$ may be replaced by the condition that $\Omega \in L^{2,1}$.

As for the function V, since it is harmonic, we can call upon Lemma 3.2 to deduce that for every $0 < \delta < 3/4$ there holds

$$\begin{aligned} \|\nabla V\|^2_{L^{2,\infty}(B_{\delta r}(p))} &\leq \|\nabla V\|^2_{L^2(B_{\delta r}(p))} \\ &\leq \left(\frac{4\delta}{3}\right)^2 \|\nabla V\|^2_{L^2(B_{3r/4}(p))} \\ &\leq C_1 \left(\frac{4\delta}{3}\right)^2 \|\nabla V\|^2_{L^{2,\infty}(B_r(p))}, \end{aligned} \tag{3.41}$$

where C_1 is a constant independent of r. Indeed, the $L^{2,\infty}$-norm of a harmonic function on the unit ball controls all its other norms on balls of radii inferior to $3/4$.

We next choose δ independent of r and so small as to have $C_1 \left(\frac{4\delta}{3}\right)^2 < 1/16$. We also adjust ε_0 to satisfy $C_0\varepsilon_0 < 1/8$. Then, combining (3.40) and (3.41) yields the following inequality

$$\|\nabla W\|_{L^{2,\infty}(B_{\delta r}(p))} \leq \frac{1}{2}\|\nabla W\|_{L^{2,\infty}(B_r(p))}, \tag{3.42}$$

valid for all $r < \rho_0$ and all $p \in B_{1/2}(0)$.

Just as in the regularity proof for the CMC equation, the latter is iterated to eventually produce the estimate

$$\sup_{p \in B_{1/2}(0)\ ,\ 0<\rho<1/4} \rho^{-\alpha}\|\nabla W\|_{L^{2,\infty}(B_\rho(p))} < +\infty. \tag{3.43}$$

Calling once again upon the duality $L^{2,1} - L^{2,\infty}$, and upon the upper bound on $\|\Omega\|_{L^{2,1}(D^2)}$ provided in (3.43), we infer that

$$\sup_{p \in B_{1/2}(0)\ ,\ 0<\rho<1/4} \rho^{-\alpha}\|\Omega \cdot \nabla W\|_{L^1(B_\rho(p))} < +\infty, \tag{3.44}$$

thereby giving (3.39). This concludes the proof of the desired statement. □

4 The regularity of critical points to quadratic, coercive and conformally invariant Lagrangians in two dimension: the proof of Hildebrandt's conjecture

The methods which we have used up to now to approach Hildebrandt's conjecture and obtain the regularity of $W^{1,2}$ solutions of the generic system

$$\Delta u + A(u)(\nabla u, \nabla u) = H(u)(\nabla^\perp u, \nabla u) \tag{4.1}$$

rely on two main ideas:

i) recast, as much as possible, quadratic nonlinear terms as linear combinations of Jacobians or as *null forms*;
ii) project equation (4.1) on a *moving frame* $(e_1 \cdots e_n)$ satisfying the *Coulomb gauge condition*

$$\forall i, j = 1 \cdots m \qquad \mathrm{div}((e_j, \nabla e_i)) = 0.$$

Both approaches can be combined to establish the Hölder continuity of $W^{1,2}$ solutions of (4.1) when the target manifold N^n is C^2, and when the prescribed mean curvature H is Lipschitz continuous (see [2,4,19]). Seemingly, these are the weakest possible hypotheses required to carry out the above strategy.

However, to fully solve Hildebrandt's conjecture, one must replace the Lipschitzean condition on H by its being an element of L^∞. This makes quite a difference!

Despite its evident elegance and verified usefulness, Hélein's moving frames method suffers from a relative opacity[18]: what makes nonlinearities of the form

$$A(u)(\nabla u, \nabla u) - H(u)(\nabla^\perp u, \nabla u)$$

so special and more favorable to treating regularity/compactness matters than seemingly simpler nonlinearities, such as

$$|\nabla u|^2,$$

which we encountered in Section 1?

The moving frames method does not address this question.

We consider a weakly harmonic map u with finite energy, on D^2 and taking values in a regular oriented closed submanifold $N^n \subset \mathbb{R}^{n+1}$ of codimension 1. We saw at the end of Section 2 that u satisfies the equation

$$-\Delta u = \nu(u)\ \nabla(\nu(u)) \cdot \nabla u, \tag{4.2}$$

where ν is the normal unit-vector to N^n relative to the orientation of N^n.

18 Yet another drawback of the moving frames method is that it lifts an N^n-valued harmonic map, with $n > 2$, to another harmonic map, valued in a parallelizable manifold $(S^1)^q$ of higher dimension. This procedure requires that N^n have a higher regularity than the "natural" one (namely, C^5 in place of C^2). It is only under this more stringent assumption that the regularity of N^n-valued harmonic maps was obtained in [3] and [19]. The introduction of Schrödinger systems with antisymmetric potentials in [27] enabled to improve these results.

In local coordinates, (4.2) may be recast as

$$-\Delta u^i = \nu(u)^i \sum_{j=1}^{n+1} \nabla(\nu(u))_j \cdot \nabla u^j \qquad \forall\ i = 1 \cdots n+1. \qquad (4.3)$$

In this more general framework, we may attempt to adapt Hélein's operation which changes (3.24) into (3.25). The first step of this process is easily accomplished. Indeed, since ∇u is orthogonal to $\nu(u)$, there holds

$$\sum_{j=1}^{n+1} \nu_j(u)\, \nabla u^j = 0.$$

Substituting this identity into (4.4) yields another equivalent formulation of the equation satisfied by N^n-valued harmonic maps, namely

$$-\Delta u^i = \sum_{j=1}^{n+1} \left(\nu(u)^i\, \nabla(\nu(u))_j - \nu(u)_j\, \nabla(\nu(u))^i\right) \cdot \nabla u^j. \qquad (4.4)$$

On the contrary, the second step of the process can not *a priori* be extended. Indeed, one cannot justify that the vector field

$$\nu(u)^i\, \nabla(\nu(u))_j - \nu(u)_j\, \nabla(\nu(u))^i$$

is divergence-free. This was true so long as N^n was the sphere S^n, but it fails so soon as the metric is ever so slightly perturbed. What remains however robust is the *antisymmetry* of the matrix

$$\Omega := \left(\nu(u)^i\, \nabla(\nu(u))_j - \nu(u)_j\, \nabla(\nu(u))^i\right)_{i,j=1\cdots n+1}. \qquad (4.5)$$

It turns out that the antisymmetry of Ω lies in the heart of the problem we have been tackling in these lecture notes. The following result sheds some light onto this claim.

Theorem 4.1 ([24]). *Let Ω be a vector field in $L^2(\wedge^1 D^2 \otimes so(m))$, thus takings values in the space antisymmetric $m \times m$ matrices $so(m)$. Suppose that u is a map in $W^{1,2}(D^2, \mathbb{R}^m)$ satisfying the equation*[19]

$$-\Delta u = \Omega \cdot \nabla u \qquad in \quad \mathcal{D}'(D^2). \qquad (4.6)$$

[19] In local coordinates, (4.6) reads

$$-\Delta u^i = \sum_{j=1}^{m} \Omega^i_j \cdot \nabla u^j \qquad \forall\ i = 1 \cdots m.$$

Then there exists some $p > 2$ such that $u \in W^{1,p}_{\mathrm{loc}}(D^2, \mathbb{R}^m)$. In particular, u is Hölder continuous. □

Prior to delving into the proof of this theorem, let us first examine some of its implications towards answering the questions we aim to solve.

First of all, it is clear that Theorem 4.1 is applicable to the equation (4.4) so as to yield the regularity of harmonic maps taking values in a manifold of codimension 1.

Another rather direct application of Theorem 4.1 deals with the solutions of the prescribed mean curvature equation in $\mathbb{R}^3$,

$$\Delta u = 2H(u)\, \partial_x u \times \partial_y u \qquad \text{in} \quad \mathcal{D}'(D^2).$$

This equation can be recast in the form

$$\Delta u = H(u) \nabla^\perp u \times \nabla u.$$

Via introducing

$$\Omega := H(u) \begin{pmatrix} 0 & -\nabla^\perp u_3 & \nabla^\perp u_2 \\ \nabla^\perp u_3 & 0 & -\nabla^\perp u_1 \\ -\nabla^\perp u_2 & \nabla^\perp u_1 & 0 \end{pmatrix}$$

we observe successively that Ω is antisymmetric, that it belongs to L^2 whenever H belongs to L^∞, and that u satisfies (4.6). The hypotheses of Theorem 4.1 are thus all satisfied, and so we conclude that that u is Hölder continuous.

This last example outlines clearly the usefulness of Theorem 4.1 towards solving Hildebrandt's conjecture. Namely, it enables us to weaken the Lipschitzean assumption on H found in previous works ([2, 16–18]), by only requiring that H be an element of L^∞. This is precisely the condition stated in Hildebrandt's conjecture. By all means, we are in good shape.

In fact, Hildebrandt's conjecture will be completely resolved with the help of the following result.

Theorem 4.2 ([24]). *Let N^n be an arbitrary closed oriented C^2-submanifold of $\mathbb{R}^m$, with $1 \leq n < m$, and let ω be a C^1 two-form on N^n. Suppose that u is a critical point in $W^{1,2}(D^2, N^n)$ of the energy*

$$E^\omega(u) = \frac{1}{2} \int_{D^2} |\nabla u|^2(x, y)\, dx\, dy + u^*\omega.$$

Then u fulfills all of the hypotheses of Theorem 4.1, and therefore is Hölder continuous.

Proof of Theorem 4.2. The critical points of E^ω satisfy the equation (2.25), which, in local coordinates, takes the form

$$\Delta u^i = -\sum_{j,k=1}^{m} H^i_{jk}(u)\,\nabla^\perp u^k\cdot\nabla u^j - \sum_{j,k=1}^{m} A^i_{jk}(u)\,\nabla u^k\cdot\nabla u^j, \qquad (4.7)$$

for $i = 1\cdots m$. Denoting by $(\varepsilon_i)_{i=1\cdots m}$ the canonical basis of $\mathbb{R}^m$, we first observe that since

$$H^i_{jk}(z) = d\omega_z(\varepsilon_i, \varepsilon_j, \varepsilon_k)$$

the antisymmetry of the 3-form $d\omega$ yields for every $z\in\mathbb{R}^m$ the identity $H^i_{jk}(z) = -H^j_{ik}(z)$. Then, (4.7) becomes

$$\Delta u^i = -\frac{1}{2}\sum_{j,k=1}^{m}(H^i_{jk}(u) - H^j_{ik}(u))\nabla^\perp u^k\cdot\nabla u^j - \sum_{j,k=1}^{m} A^i_{jk}(u)\nabla u^k\cdot\nabla u^j. \quad (4.8)$$

On the other hand, $A(u)(U,V)$ is orthogonal to the tangent plane for every choice of vectors U and V[20]. In particular, there holds

$$\sum_{j=1}^{m} A^j_{ik}\,\nabla u^j = 0 \qquad \forall\ i,k = 1\cdots m. \qquad (4.9)$$

Inserting this identity into (4.8) produces

$$\begin{aligned}\Delta u^i &= -\sum_{j,k=1}^{m}(H^i_{jk}(u) - H^j_{ik}(u))\,\nabla^\perp u^k\cdot\nabla u^j\\ &\quad -\sum_{j,k=1}^{m}(A^i_{jk}(u) - A^j_{ik}(u))\,\nabla u^k\cdot\nabla u^j.\end{aligned} \qquad (4.10)$$

The $m\times m$ matrix $\Omega := (\Omega^i_j)_{i,j=1\cdots m}$ defined via

$$\Omega^i_j := \sum_{k=1}^{m}(H^i_{jk}(u) - H^j_{ik}(u))\,\nabla^\perp u^k + \sum_{k=1}^{m}(A^i_{jk}(u) - A^j_{ik}(u))\,\nabla u^k,$$

is evidently antisymmetric, and it belongs to L^2. With this notation, (4.10) is recast in the form (4.6), and thus all of the hypotheses of Theorem 4.1 are fulfilled, thereby concluding the proof of Theorem 4.2. □

[20] Rigorously speaking, A is only defined for pairs of vectors which are tangent to the surface. Nevertheless, A can be extended to all pairs of vectors in $\mathbb{R}^m$ in a neighborhood of N^n by applying the pull-back of the projection on N^n. This extension procedure is analogous to that outlined on page 130.

On the conservation laws for Schrödinger systems with antisymmetric potentials

Per the above discussion, there only remains to establish Theorem 4.1 in order to reach our goal. To this end, we will express the Schrödinger systems with antisymmetric potentials in the form of *conservation laws.* More precisely, we have

Theorem 4.3 ([24]). *Let Ω be a matrix-valued vector field on D^2 in $L^2(\wedge^1 D^2, so(m))$. Suppose that A and B are two $W^{1,2}$ functions on D^2 taking their values in the same of square $m \times m$ matrices which satisfy the equation*

$$\nabla A - A\Omega = -\nabla^{\perp} B. \tag{4.11}$$

If A is almost everywhere invertible, and if it has the bound

$$\|A\|_{L^\infty(D^2)} + \|A^{-1}\|_{L^\infty(D^2)} < +\infty, \tag{4.12}$$

then u is a solution of the Schrödinger system (4.6) *if and only if it satisfies the conservation law*

$$\operatorname{div}(A\nabla u - B\nabla^{\perp} u) = 0. \tag{4.13}$$

If (4.13) *holds, then $u \in W^{1,p}_{\text{loc}}(D^2, \mathbb{R}^m)$ for any $1 \leq p < +\infty$, and therefore u is Hölder continuous in the interior of D^2, $C^{0,\alpha}_{\text{loc}}(D^2)$ for any $\alpha < 1$.*

We note that the conservation law (4.13), when it exists, generalizes the conservation laws previously encountered in the study of problems with symmetry, namely:

1) In the case of the constant mean curvature equation, the conservation law (3.1) is (4.13) with the choice

$$A_{ij} = \delta_{ij},$$

and

$$B = \begin{pmatrix} 0 & -H\,u_3 & H\,u_2 \\ H\,u_3 & 0 & -H\,u_1 \\ -H\,u_2 & H\,u_1 & 0 \end{pmatrix}.$$

2) In the case of S^n-valued harmonic maps, the conservation law (3.25) is (4.13) for

$$A_{ij} = \delta_{ij}$$

and $B = (B^i_j)$ with

$$\nabla^{\perp} B^i_j = u^i\,\nabla u_j - u_j\,\nabla u^i.$$

The ultimate part of this section will be devoted to constructing A and B, for any given antisymmetric Ω, with sufficiently small L^2-norms (*cf.* Theorem 4.5 below). As a result, all coercive conformally invariant Lagrangians with quadratic growth will yield conservation laws written in divergence form. This is quite an amazing fact. Indeed, while in cases of the CMC and S^n-valued harmonic map equations the existence of conservation laws can be explained by Noether's theorem[21], one may wonder **which hidden symmetries yield the existence of the general divergence form (4.13)?** This profound question shall unfortunately not be addressed here.

Prior to constructing A and B in the general case, we first establish Theorem 4.3.

Proof of Theorem 4.3. The first part of the theorem is the result of the elementary calculation,

$$\begin{aligned}\operatorname{div}(A\,\nabla u - B\,\nabla^{\perp} u) &= A\,\Delta u + \nabla A \cdot \nabla u - \nabla B \cdot \nabla^{\perp} u\\ &= A\,\Delta u + (\nabla A + \nabla^{\perp} B)\cdot \nabla u\\ &= A(\Delta u + \Omega \cdot \nabla u) = 0.\end{aligned}$$

Regularity matters are settled as follows. Just as in the previously encountered problems, we seek to employ a Morrey-type argument via the existence of some constant $\alpha > 0$ such that

$$\sup_{p\in B_{1/2}(0)\,,\,0<\rho<1/4} \rho^{-\alpha} \int_{B_\rho(p)} |\Delta u| < +\infty. \tag{4.14}$$

The fact that ∇u belongs to $L^p_{\mathrm{loc}}(D^2)$ for some $p > 2$ is then deduced through calling upon the inequalities in [1], exactly in the same manner as we previously outlined. Finally once we know that ∇u belongs to $L^p_{\mathrm{loc}}(D^2)$ for some $p > 2$ we deduce the whole regularity result stated in the theorem by using the following lemma which can be proved using classical arguments.

Lemma 4.4. *Let* $m \in \mathbb{N}\setminus\{0\}$ *and* $u \in W^{1,p}_{\mathrm{loc}}(D^2,\mathbb{R}^m)$ *for some* $p > 2$ *satisfying*

$$-\Delta u = \Omega \cdot \nabla u$$

[21] Roughly speaking, symmetries give rise to conservation laws. In both the CMC and S^n-harmonic map equations, the said symmetries are tantamount to the corresponding Lagrangians being invariant under the group of isometries of the target space $\mathbb{R}^m$.

where[22] $\Omega \in L^2(D^2, M_m(\mathbb{R}) \otimes \mathbb{R}^2)$ *then* $u \in W^{1,q}_{\mathrm{loc}}(D^2, \mathbb{R}^m)$ *for any* $q < +\infty$.

Let $\varepsilon_0 > 0$ be some constant whose value will be adjusted in due time to fit our needs. There exists a radius ρ_0 such that for every $r < \rho_0$ and every point p in $B_{1/2}(0)$, there holds

$$\int_{B_r(p)} |\nabla A|^2 + |\nabla B|^2 < \varepsilon_0. \tag{4.15}$$

Henceforth, we consider only radii $r < \rho_0$.

Note that $A\nabla u$ satisfies the elliptic system

$$\begin{cases} \operatorname{div}(A\nabla u) = \nabla B \cdot \nabla^\perp u = \partial_y B\, \partial_x u - \partial_x B\, \partial_y u \\ \operatorname{rot}(A\nabla u) = -\nabla A \cdot \nabla^\perp u = \partial_x A\, \partial_y u - \partial_y A\, \partial_x u. \end{cases}$$

We proceed by introducing on $B_r(p)$ the linear Hodge decomposition in L^2 of $A\nabla u$. Namely, there exist two functions C and D, unique up to additive constants, elements of $W_0^{1,2}(B_r(p))$ and $W^{1,2}(B_r(p))$ respectively, and such that

$$A\nabla u = \nabla C + \nabla^\perp D. \tag{4.16}$$

To see why such C and D do indeed exist, consider first the equation

$$\begin{cases} \Delta C = \operatorname{div}(A\nabla u) = \partial_y B\, \partial_x u - \partial_x B\, \partial_y u & \text{in } B_r(p) \\ C = 0 & \text{on } \partial B_r(p). \end{cases} \tag{4.17}$$

Wente's Theorem (3.1) guarantees that C lies in $W^{1,2}$, and moreover

$$\int_{B_r(p)} |\nabla C|^2 \le C_0 \int_{B_r(p)} |\nabla B|^2 \int_{B_r(p)} |\nabla u|^2. \tag{4.18}$$

By construction, $\operatorname{div}(A\nabla u - \nabla C) = 0$. Poincaré's lemma thus yields the existence of D in $W^{1,2}$ with $\nabla^\perp D := A\nabla u - \nabla C$, and

$$\begin{aligned} \int_{B_r(p)} |\nabla D|^2 &\le 2\int_{B_r(p)} |A\nabla u|^2 + |\nabla C|^2 \\ &\le 2\|A\|_\infty \int_{B_r(p)} |\nabla u|^2 + 2C_0 \int_{B_r(p)} |\nabla B|^2 \int_{B_r(p)} |\nabla u|^2. \end{aligned} \tag{4.19}$$

[22] Observe that in this lemma no antisymmetry assumption is made for Ω which is an arbitrary $m \times m$–matrix valued L^2–vectorfield on D^2.

The function D satisfies the identity

$$\Delta D = -\nabla A \cdot \nabla^{\perp} u = \partial_x A\, \partial_y u - \partial_y A\, \partial_x u.$$

Just as we did in the case of the CMC equation, we introduce the decomposition $D = \phi + v$, with ϕ fulfilling

$$\begin{cases} \Delta\phi = \partial_x A\, \partial_y u - \partial_y A\, \partial_x u & \text{in} \quad B_r(p) \\ \phi = 0 & \text{on} \quad \partial B_r(p) \end{cases} \tag{4.20}$$

and with v being harmonic. Once again, Wente's Theorem 3.1 gives us the estimate

$$\int_{B_r(p)} |\nabla\phi|^2 \le C_0 \int_{B_r(p)} |\nabla A|^2 \int_{B_r(p)} |\nabla u|^2. \tag{4.21}$$

The arguments which we used in the course of the regularity proof for the CMC equation may be recycled here so as to obtain the analogous version of (3.16), only this time on the ball $B_{\delta r}(p)$, where $0 < \delta < 1$ will be adjusted in due time. More precisely, we find

$$\begin{aligned} \int_{B_{\delta r}(p)} |\nabla D|^2 \le 2\delta^2 \int_{B_r(p)} |\nabla D|^2 \\ +3 \int_{B_r(p)} |\nabla\phi|^2. \end{aligned} \tag{4.22}$$

Bringing altogether (4.15), (4.18), (4.19), (4.21) and (4.22) produces

$$\begin{aligned} \int_{B_{\delta r}(p)} |A\, \nabla u|^2 \le 3\delta^2 \int_{B_r(p)} |A\, \nabla u|^2 \\ +C_1\, \varepsilon_0 \int_{B_r(p)} |\nabla u|^2. \end{aligned} \tag{4.23}$$

Using the hypotheses that A and A^{-1} are bounded in L^∞, it follows from (4.23) that for all $0 < \delta < 1$, there holds the estimate

$$\begin{aligned} \int_{B_{\delta r}(p)} |\nabla u|^2 \le 3\|A^{-1}\|_\infty \|A\|_\infty \delta^2 \int_{B_r(p)} |\nabla u|^2 \\ +C_1 \|A^{-1}\|_\infty \varepsilon_0 \int_{B_r(p)} |\nabla u|^2. \end{aligned} \tag{4.24}$$

Next, we choose ε_0 and δ strictly positive, independent of r and p, and such that

$$3\|A^{-1}\|_\infty \|A\|_\infty \delta^2 + C_1 \|A^{-1}\|_\infty \varepsilon_0 = \frac{1}{2}.$$

For this particular choice of δ, we have thus obtained the inequality

$$\int_{B_{\delta r}(p)} |\nabla u|^2 \leq \frac{1}{2} \int_{B_r(p)} |\nabla u|^2.$$

Iterating this inequality as in the previous regularity proofs yields the existence of some constant $\alpha > 0$ for which

$$\sup_{p \in B_{1/2}(0)\ ,\ 0<\rho<1/4} \rho^{-2\alpha} \int_{B_\rho(p)} |\nabla u|^2 < +\infty.$$

Since $|\Delta u| \leq |\Omega|\, |\nabla u|$, the latter gives us (4.14), thereby concluding the proof of Theorem 4.3. □

There only now remains to establish the existence of the functions A and B in $W^{1,2}$ satisfying the equation (4.11) and the hypothesis (4.12).

The construction of conservation laws for systems with antisymmetric potentials, and the proof of Theorem 4.1

The following result, combined to Theorem 4.3, implies Theorem 4.1, itself yielding Theorem 4.2, and thereby providing a proof of Hildebrandt's conjecture, as we previously explained.

Theorem 4.5 ([24]). *There exists a constant $\varepsilon_0(m) > 0$ depending only on the integer m, such that for every vector field $\Omega \in L^2(D^2, so(m))$ with*

$$\int_{D^2} |\Omega|^2 < \varepsilon_0(m), \tag{4.25}$$

it is possible to construct $A \in L^\infty(D^2, Gl_m(\mathbb{R})) \cap W^{1,2}$ and $B \in W^{1,2}(D^2, M_m(\mathbb{R}))$ with the properties

i) $$\int_{D^2} |\nabla A|^2 + \|\mathrm{dist}(A, SO(m))\|_{L^\infty(D^2)} \leq C(m) \int_{D^2} |\Omega|^2, \tag{4.26}$$

ii) $$\nabla_\Omega A := \mathrm{div}\nabla A - A\Omega = -\nabla^\perp B, \tag{4.27}$$

where $C(m)$ is a constant depending only on the dimension m.

Prior to delving into the proof of Theorem 4.5, a few comments and observations are in order.

Glancing at the statement of the theorem, one question naturally arises: **why is the antisymmetry of Ω so important?**

It can be understood as follow.

In the simpler case when Ω is **divergence-free**, we can write Ω in the form

$$\Omega = \nabla^{\perp}\xi,$$

for some $\xi \in W^{1,2}(D^2, so(m))$. In particular, the statement of Theorem 4.5 is settled by choosing

$$A_{ij} = \delta_{ij} \qquad \text{and} \qquad B_{ij} = \xi_{ij}. \tag{4.28}$$

Accordingly, it seems reasonable in the general case to seek a solution pair (A, B) which comes as "close" as can be to (4.28). A first approach consists in performing a **linear Hodge decomposition** in L^2 of Ω. Hence, for some ξ and P in $W^{1,2}$, we write

$$\Omega = \nabla^{\perp}\xi - \nabla P. \tag{4.29}$$

In this case, we see that if A exists, then it must satisfy the equation

$$\Delta A = \nabla A \cdot \nabla^{\perp}\xi - \mathrm{div}(A\nabla P). \tag{4.30}$$

This equation is critical in $W^{1,2}$. The first summand $\nabla A \cdot \nabla^{\perp}\xi$ on the right-hand side of (4.30) is a Jacobian. This is a desirable feature with many very good analytical properties, as we have previously seen. In particular, using integration by compensation (Wente's Theorem 3.1), we can devise a bootstrapping argument beginning in $W^{1,2}$. On the other hand, the second summand $\mathrm{div}(A\nabla P)$ on the right-hand side of (4.30) displays no particular structure. All that we know about it, is that A should *a-priori* belong to $W^{1,2}$. But this space is not embedded in L^∞, and so we cannot a priori conclude that $A\nabla P$ lies in L^2, thereby obstructing a successful analysis...

However, not all hope is lost for the **antisymmetric structure** of Ω still remains to be used. The idea is to perform a **nonlinear** Hodge decomposition[23] in L^2 of Ω. Thus, let $\xi \in W^{1,2}(D^2, so(m))$ and P be a $W^{1,2}$ map taking values in the group $SO(m)$ of proper rotations of $\mathbb{R}^m$, such that

$$\Omega = P\,\nabla^{\perp}\xi\, P^{-1} - \nabla P\, P^{-1}. \tag{4.31}$$

At first glance, the advantage of (4.31) over (4.30) is not obvious. If anything, it seems as though we have complicated the problem by having to introduce left and right multiplications by P and P^{-1}. On second thought, however, since rotations are always bounded, the map P in

[23] Which is tantamount to a **change of gauge**.

(4.31) is an element of $W^{1,2} \cap L^\infty$, whereas in (4.30), the map P belonged only to $W^{1,2}$. This slight improvement will actually be sufficient to successfully carry out our proof. Furthermore, (4.31) has yet another advantage over (4.30). Indeed, whenever A and B are solutions of (4.27), there holds

$$\begin{aligned}\nabla_{\nabla^\perp\xi}(AP) &= \nabla(AP) - (AP)\,\nabla^\perp\xi\\ &= \nabla A\,P + A\,\nabla P - A\,P\,(P^{-1}\Omega P + P^{-1}\nabla P)\\ &= (\nabla_\Omega A)P = -\nabla^\perp B\;P.\end{aligned}$$

Hence, via setting $\tilde{A} := AP$, we find

$$\Delta\tilde{A} = \nabla\tilde{A}\cdot\nabla^\perp\xi + \nabla^\perp B\cdot\nabla P. \tag{4.32}$$

Unlike (4.30), the second summand on the right-hand side of (4.32) is a linear combination of Jacobians of terms which lie in $W^{1,2}$. Accordingly, calling upon Theorem 3.1, we can control $\tilde{A}$ in $L^\infty \cap W^{1,2}$. This will make a bootstrapping argument possible.

One point still remains to be verified. Namely, that the nonlinear Hodge decomposition (4.31) does exist. This can be accomplished with the help of a result of Karen Uhlenbeck[24].

Theorem 4.6 ([24,35]). *Let $m \in \mathbb{N}$. There are two constants $\varepsilon(m) > 0$ and $C(m) > 0$, depending only on m, such that for each vector field $\Omega \in L^2(D^2, so(m))$ with*

$$\int_{D^2}|\Omega|^2 < \varepsilon(m),$$

there exist $\xi \in W^{1,2}(D^2, so(m))$ and $P \in W^{1,2}(D^2, SO(m))$ satisfying

$$\Omega = P\nabla^\perp\xi P^{-1} - \nabla P P^{-1}, \tag{4.33}$$

$$\xi = 0 \quad on \quad \partial D^2 \tag{4.34}$$

and

$$\int_{D^2}|\nabla\xi|^2 + \int_{D^2}|\nabla P|^2 \le C(m)\int_{D^2}|\Omega|^2. \tag{4.35}$$

[24] In reality, this result, as it is stated here, does not appear in the original work of Uhlenbeck. In [24], it is shown how to deduce Theorem 4.6 from Uhlenbeck's approach.

Proof of Theorem 4.5. Let P and ξ be as in Theorem 4.6. To each $A \in L^\infty \cap W^{1,2}(D^2, M_m(\mathbb{R}))$ we associate $\tilde{A} = AP$. Suppose that A and B are solutions of (4.27). Then $\tilde{A}$ and B satisfy the elliptic system

$$\begin{cases} \Delta \tilde{A} = \nabla \tilde{A} \cdot \nabla^\perp \xi + \nabla^\perp B \cdot \nabla P \\ \Delta B = -\mathrm{div}(\tilde{A}\, \nabla \xi \; P^{-1}) + \nabla^\perp \tilde{A} \cdot \nabla P^{-1}. \end{cases} \tag{4.36}$$

We first consider the invertible elliptic system

$$\begin{cases} \Delta \tilde{A} = \nabla \hat{A} \cdot \nabla^\perp \xi + \nabla^\perp \hat{B} \cdot \nabla P \\ \Delta B = -\mathrm{div}(\hat{A}\, \nabla \xi \; P^{-1}) + \nabla^\perp \hat{A} \cdot \nabla P^{-1} \\ \dfrac{\partial \tilde{A}}{\partial \nu} = 0 \quad \text{and} \quad B = 0 \quad \text{on} \quad \partial D^2 \\ \displaystyle\int_{D^2} \tilde{A} = \pi^2 \, Id_m \end{cases} \tag{4.37}$$

where $\hat{A}$ and $\hat{B}$ are arbitrary functions in $L^\infty \cap W^{1,2}$ and in $W^{1,2}$ respectively. An analogous version[25] of Theorem 3.1 with Neumann boundary conditions in place of Dirichlet conditions, we deduce that the unique solution $(\tilde{A}, B)$ of (4.36) satisfies the estimates

$$\begin{aligned} \int_{D^2} |\nabla \tilde{A}|^2 + \|\tilde{A} - Id_m\|_\infty^2 \leq C \int_{D^2} |\nabla \hat{A}|^2 \int_{D^2} |\nabla \xi|^2 \\ + C \int_{D^2} |\nabla \hat{B}|^2 \int_{D^2} |\nabla P|^2, \end{aligned} \tag{4.38}$$

and

$$\begin{aligned} \int_{D^2} |\nabla(\tilde{B} - B_0)|^2 \leq C \|\hat{A} - Id_m\|_\infty^2 \int_{D^2} |\nabla \xi|^2 \\ + C \int_{D^2} |\nabla \hat{A}|^2 \int_{D^2} |\nabla P|^2, \end{aligned} \tag{4.39}$$

where B_0 is the solution in $W^{1,2}$ of

$$\begin{cases} \Delta B_0 = -\mathrm{div}(\nabla \xi \; P^{-1}) & \text{in } D^2 \\ B_0 = 0 & \text{on } \partial D^2. \end{cases} \tag{4.40}$$

[25] Whose proof is left as an exercise.

Hence, if

$$\int_{D^2} |\nabla P|^2 + |\nabla \xi|^2$$

is sufficiently small (this can always be arranged owing to (4.35) and the hypothesis (4.25)), then a standard fixed point argument in the space $\left(L^\infty \cap W^{1,2}(D^2, M_m(\mathbb{R}))\right) \times W^{1,2}(D^2, M_m(\mathbb{R}))$ yields the existence of the solution $(\tilde{A}, B)$ of the system

$$\begin{cases} \Delta \tilde{A} = \nabla \tilde{A} \cdot \nabla^\perp \xi + \nabla^\perp B \cdot \nabla P \\ \Delta B = -\mathrm{div}(\tilde{A}\, \nabla \xi\ P^{-1}) + \nabla^\perp \tilde{A} \cdot \nabla P^{-1} \\ \dfrac{\partial \tilde{A}}{\partial \nu} = 0 \quad \text{and} \quad B = 0 \quad \text{on} \quad \partial D^2 \\ \displaystyle\int_{D^2} \tilde{A} = \pi^2\, Id_m. \end{cases} \tag{4.41}$$

By construction, this solution satisfies the estimate (4.26) with $A = \tilde{A}\, P^{-1}$.

The proof of Theorem 4.5 will then be finished once it is established that (A, B) is a solution of (4.27).

To do so, we introduce the following linear Hodge decomposition in L^2:

$$\nabla \tilde{A} - \tilde{A} \nabla^\perp \xi + \nabla^\perp B\ P = \nabla C + \nabla^\perp D$$

where $C = 0$ on ∂D^2. The first equation in (4.41) states that $\Delta C = 0$, so that $C \equiv 0$ in D^2. The second equation in (4.41) along with the boundary conditions implies that D satisfies

$$\begin{cases} \mathrm{div}(\nabla D\ P^{-1}) = 0 & \text{in } D^2 \\ D = 0 & \text{on } \partial D^2. \end{cases} \tag{4.42}$$

Thus, there exists $E \in W^{1,2}(D^2, M_n(\mathbb{R}))$ such that

$$\begin{cases} -\Delta E = \nabla^\perp D \cdot \nabla P^{-1} & \text{in } D^2 \\ \dfrac{\partial E}{\partial \nu} = 0 & \text{on } \partial D^2. \end{cases} \tag{4.43}$$

The analogous version of Theorem 3.1 with Neumann boundary conditions yields the estimate

$$\int_{D^2} |\nabla E| \le C_0 \int_{D^2} |\nabla D|^2 \int_{D^2} |\nabla P^{-1}|^2. \tag{4.44}$$

Moreover, because $\nabla D = \nabla^\perp E\ P$, there holds $|\nabla D| \leq |\nabla E|$. Put into (4.44), this shows that if $\int_{D^2} |\nabla P|^2$ is chosen sufficiently small (*i.e.* for $\varepsilon_0(m)$ in (4.25) small enough), then $D \equiv 0$. Whence, we find

$$\nabla \tilde{A} - \tilde{A}\nabla^\perp \xi + \nabla^\perp B P = 0 \qquad \text{in} \quad D^2,$$

thereby ending the proof of Theorem 4.5. □

5 A PDE version of the constant variation method for Schrödinger Systems with anti-symmetric potentials

In this part we shall look at various, a-priori critical, elliptic systems with antisymmetric potentials and extend the approach we developed in the previous section in order to establish their hidden sub-critical nature. Before to do so we will look at what we have done so far from a different perspective than the one suggested by the "gauge theoretic" type arguments we used.

In 2 dimension, in order to prove the sub-criticality of the system

$$-\Delta u = \Omega \cdot \nabla u, \tag{5.1}$$

where $u \in W^{1,2}(D^2, \mathbb{R}^m)$ and $\Omega \in L^2(\wedge^1 D^2, so(m))$ – sub-criticality meaning that the fact that u solves (5.1) imply that it is in fact more regular than the initial assumption $u \in W^{1,2}$ – we proceeded as follows: we first constructed a solution of the equation

$$\mathrm{div}(\nabla P P^{-1}) = \mathrm{div}(P \Omega P^{-1}), \tag{5.2}$$

and multiplying ∇u by the rotation P in the equation (5.1) we obtain that it is equivalent to

$$-\mathrm{div}(P\, \nabla u) = \nabla^\perp \xi \cdot P\, \nabla u, \tag{5.3}$$

where

$$\nabla^\perp \xi := -\nabla P\, P^{-1} + P\, \Omega\, P^{-1}.$$

Then in order to have a pure Jacobian in the right hand side to (5.3) we looked for a replacement of P by a perturbation of the form $A := (Id_m + \varepsilon)\ P$ where ε is a $m \times m$ matrix valued map hopefully small in $L^\infty \cap W^{1,2}$. This is obtained by solving the following well posed problem in $W^{1,2} \cap L^\infty$ – due to Wente's theorem –

$$\begin{cases} \mathrm{div}(\nabla \varepsilon P) = \mathrm{div}((Id_m + \varepsilon)\ \nabla^\perp \xi\ P) & \text{in } D^2 \\ \varepsilon = 0 & \text{on } \partial D^2. \end{cases} \tag{5.4}$$

Posing $\nabla^\perp B := \nabla\varepsilon\, P - (Id_m + \varepsilon)\, \nabla^\perp \xi\, P$ equation (5.1) becomes equivalent to

$$-\mathrm{div}(A\, \nabla u) = \nabla^\perp B \cdot \nabla u. \tag{5.5}$$

Trying now to reproduce this procedure in one space dimension leads to the following. We aim to solve

$$-u'' = \Omega\, u' \tag{5.6}$$

where $u : [0,1] \to \mathbb{R}^m$ and $\Omega : [0,1] \to so(m)$. We then construct a solution P to (5.3) which in one dimension becomes

$$(P' P^{-1})' = (P\, \Omega\, P^{-1})'.$$

A special solution is given by P solving

$$P^{-1}\, P' = \Omega. \tag{5.7}$$

Computing now $(Pu')'$ gives the 1-D analogue of (5.3) which is

$$(Pu')' = 0, \tag{5.8}$$

indeed the curl operator in one dimension is trivial, and jacobians too, since curl-like vector fields corresponds to functions f satisfying $\mathrm{div}\, f = f' = 0$ and then can be taken equal to zero. At this stage it is not necessary to go to the ultimate step and perturb P into $(Id_m + \varepsilon)P$ since the right hand side of (5.8) is already a pure 1-D jacobian (that is zero !) and since we have succeeded in writing equation (5.6) in conservative form.

The reader has then noticed that the 1-D analogue of our approach to write equation (5.1) in conservative form is the well known **variation of the constant method**: a solution is constructed to some **auxiliary equation** – (5.2) or (5.7) – which has been carefully chosen in order to absorb the “worst” part of the right hand side of the original equation while comparing our given solution to the constructed solution of the auxiliary equation.

We shall then in the sequel forget the geometrical interpretation of the auxiliary equation (4.33) in terms of gauge theory that we see as being specific to the kind of equation (5.1) we were looking at and keep the general philosophy of the classical variation of the constant method for Ordinary Differential Equations that we are now extending to other classes of Partial Differential Equations different from (5.1).

We establish the following result.

Theorem 5.1 ([26]). *Let $n > 2$ and $m \geq 2$; there exists $\varepsilon_0 > 0$ and $C > 0$ such that for any $\Omega \in L^{n/2}(B^n, so(m))$ there exists $A \in L^\infty \cap W^{2,n/2}(B^n, Gl_m(\mathbb{R}))$ satisfying*

i)
$$\|A\|_{W^{2,n/2}(B^n)} \leq C\ \|\Omega\|_{L^{n/2}(B^n)}, \tag{5.9}$$

ii)
$$\Delta A + A\Omega = 0. \tag{5.10}$$

Moreover for any map v in $L^{n/(n-2)}(B^n, \mathbb{R}^m)$ and $\Delta v \in L^1(B^n, \mathbb{R}^m)$

$$-\Delta v = \Omega\, v \Longleftrightarrow \operatorname{div}(A\,\nabla v - \nabla A\, v) = 0 \tag{5.11}$$

and we deduce that $v \in L^\infty_{\text{loc}}(B^n)$.

Remark 5.2. Again the assumptions $v \in L^{n/(n-2)}(B^n, \mathbb{R}^m)$ and $\Omega \in L^{n/2}(B^n, so(m))$ make equation (5.11) critical in dimension n: Inserting this information in the right hand side of (5.11) gives $\Delta v \in L^1$ which implies in return $v \in L^{n/(n-2),\infty}_{\text{loc}}$, which corresponds to our definition of being critical for an elliptic system.

Remark 5.3. We have then been able to write critical systems of the kind $-\Delta v = \Omega\, v$ in conservative form whenever Ω is antisymmetric. This "factorization of the divergence" operator is obtain through the construction of a solution A to the auxiliary equation $\Delta A + A\,\Omega = 0$ exactly like in the **constant variation method in 1-D**, a solution to the auxiliary equation $A'' + A\,\Omega = 0$ permits to factorize the derivative in the ODE given by $-z'' = \Omega\, z$ which becomes, after multiplication by A: $(A\, z' - A'\, z)' = 0$.

Proof of Theorem 5.1 *for $n = 4$.* The goal again here, like in the previous sections, is to establish a Morrey type estimate for v that could be re-injected in the equation and converted into an L^q_{loc} due to Adams result in [1].

We shall look for some map P from B^4 into the space $SO(m)$ solving some ad-hoc auxiliary equation. Formal computation – we still don't kow which regularity for P we should assume at this stage – gives

$$\begin{aligned} -\Delta(P\, v) &= -\Delta P\, v - P\Delta v - 2\nabla P \cdot \nabla v \\ &= \left(\Delta P\, P^{-1} + P\,\Omega\, P^{-1}\right)\, Pv \\ &\quad - 2\operatorname{div}(\nabla P\, P^{-1}\, Pv). \end{aligned} \tag{5.12}$$

In view of the first term in the right hand side of equation (5.12) it is natural to look for ΔP having the same regularity as Ω that is L^2. Hence we are looking for $P \in W^{2,2}(B^4, SO(m))$. Under such an assumption the second term in the right hand side is not problematic while working in the function space L^2 for v indeed, standard elliptic estimates give

$$\|\Delta_0^{-1}\left(\mathrm{div}(\nabla P\, P^{-1}\, Pv)\right)\|_{L^2(B^4)} \leq \|\nabla P\|_{L^4(B^4)}\, \|v\|_{L^2(B^4)} \qquad (5.13)$$

where Δ_0^{-1} is the map which to f in $W^{-1,4/3}$ assigns the function u in $W^{1,4/3}(B^4)$ satisfying $\Delta u = f$ and equal to zero on the boundary of B^4. Hence, if we localize in space that ensures that $\|\nabla P\|_{L^4(B^4)}$ is small enough, the contribution of the second term of the right hand side of (5.12) can be "absorbed" in the left hand side of (5.12) while working with the L^2 norm of v. □

The first term in the right hand side of (5.12) is however problematic while intending to work with the L^2 norm of v. It is indeed only in L^1 and such an estimate does not give in return an L^2 control of v. It is then tempting to look for a map P solving an auxiliary equation in such a way that this term vanishes. Unfortunately such a hope cannot be realized due to the fact that when P is a map into the rotations

$$P\,\Omega\,P^{-1} \in so(m) \quad \text{but a-priori} \quad \Delta P\,P^{-1} \notin so(m).$$

The idea is then to cancel "as much as we can" in the first term of the right hand side of (5.12) by looking at a solution to the following auxiliary equation[26]:

$$A\,\mathrm{Sym}(\Delta P P^{-1}) + P\,\Omega\,P^{-1} = 0 \qquad (5.14)$$

where $A\,\mathrm{Sym}(\Delta P P^{-1})$ is the antisymmetric part of $\Delta P P^{-1}$ given by

$$A\,\mathrm{Sym}(\Delta P P^{-1}) := \frac{1}{2}\left(\Delta P\,P^{-1} - P\,\Delta P^{-1}\right).$$

Precisely the following proposition holds

Proposition 5.4. *There exists $\varepsilon_0 > 0$ and $C > 0$ such that for any $\Omega \in L^2(B^4, so(m))$ satisfying $\|\Omega\|_{L^2} < \varepsilon_0$ there exists $P \in W^{2,2}(B^4, SO(m))$ such that*

$$A\,\mathrm{Sym}(\Delta P\,P^{-1}) + P\,\Omega\,P^{-1} = 0$$

and

$$\|\nabla P\|_{W^{1,2}(B^4)} \leq C\,\|\Omega\|_{L^2(B^4)}. \qquad (5.15)$$

[26] Which does not have, to our knowledge, a geometric relevant interpretation similar to the Coulomb Gauge extraction we used in the previous section.

Taking the gauge P given by the previous proposition and denoting $w := P\,v$ the system (5.12) becomes

$$\mathcal{L}_P w := -\Delta w - (\nabla P\; P^{-1})^2\; w + 2\,\mathrm{div}(\nabla P\; P^{-1}\; w) = 0, \qquad (5.16)$$

where we have used that

$$\begin{aligned}\mathrm{Symm}(\Delta P P^{-1}) &= 2^{-1}\left(\Delta P P^{-1} - P\,\Delta P^{-1}\right)\\ &= 2^{-1}\mathrm{div}\left(\nabla P P^{-1} + P\nabla P^{-1}\right) - \nabla P\cdot\nabla P^{-1}\\ &= -(\nabla P P^{-1})^2.\end{aligned}$$

Denote for any $Q \in W^{2,2}(B^4, M_m(\mathbb{R}))$ $\mathcal{L}_P^* Q$ the "formal adjoint" to $\mathcal{L}_P$ acting on Q

$$\mathcal{L}_P^* Q := -\Delta Q - 2\,\nabla Q\cdot\nabla P\; P^{-1} - Q\,(\nabla P\; P^{-1})^2.$$

In order to factorize the divergence operator in (5.16) it is natural to look for Q satisfying $\mathcal{L}_P^* Q = 0$. One has indeed

$$\begin{aligned}0 &= w\mathcal{L}_P^* Q - Q\mathcal{L}_P w\\ &= \mathrm{div}(Q\nabla w - [\nabla Q + 2Q\nabla P\; P^{-1}]\; w).\end{aligned} \qquad (5.17)$$

It is not so difficult to construct Q solving $\mathcal{L}_P^* Q = 0$ for a $Q \in W^{2,p}(B^4, M_m(\mathbb{R}))$ (for $p < 2$). However, in order to give a meaning to (5.17) we need at least $Q \in W^{2,2}$ and, moreover the invertibility of the matrix Q almost everywhere is also needed for the conservation law. In the aim of producing a $Q \in W^{2,2}(B^4, Gl_m(\mathbb{R}))$ solving $\mathcal{L}_P^* Q = 0$ it is important to observe first that $-(\nabla P\; P^{-1})^2$ is a non-negative symmetric matrix[27] but that it is in a smaller space than L^2: the space $L^{2,1}$. Combining the improved Sobolev embeddings[28] space $W^{1,2}(B^4) \hookrightarrow L^{4,2}(B^4)$ and the fact that the product of two functions in $L^{4,2}$ is in $L^{2,1}$ (see [33]) we deduce from (5.15) that

$$\|(\nabla P P^{-1})^2\|_{L^{2,1}(B^4)} \leq C\|\Omega\|^2_{L^2(B^4)}. \qquad (5.18)$$

Granting these three important properties for $-(\nabla P P^{-1})^2$ (symmetry, positiveness and improved integrability) one can prove the following result (see [26]).

---

27 Indeed it is a sum of non-negative symmetric matrices: each of the matrices $\partial_{x_j} P\; P^{-1}$ is anti-symmetric and hence its square $(\partial_{x_j} P\; P^{-1})^2$ is symmetric non-positive.

28 $L^{4,2}(B^4)$ is the Lorentz space of measurable functions f such that the decreasing rearrangement f^* of f satisfies $\int_0^\infty t^{-1/2}(f^*)^2(t)\,dt < +\infty$.

Theorem 5.5. *There exists* $\varepsilon > 0$ *such that*

$$\forall P \in W^{2,2}(B^4, SO(m)) \quad \textit{satisfying} \quad \|\nabla P\|_{W^{1,2}(B^4)} \leq \varepsilon$$

there exists a unique $Q \in W^{2,2} \cap L^\infty(B^4, Gl_m(\mathbb{R}))$ *satisfying*

$$\begin{cases} -\Delta Q - 2\,\nabla Q \cdot \nabla P\, P^{-1} - Q\,(\nabla P\, P^{-1})^2 = 0 \\ Q = Id_m & \textit{on } \partial B^4 \end{cases} \tag{5.19}$$

and

$$\|Q - Id_m\|_{L^\infty \cap W^{2,2}} \leq C\,\|\nabla P\|^2_{W^{1,2}(B^4)}. \tag{5.20}$$

Taking $A := P\,Q$ we have constructed a solution to

$$\Delta A + A\Omega = 0 \tag{5.21}$$

such that

$$\|\mathrm{dist}(A, SO(m))\|_\infty + \|A - Id\|_{W^{2,2}} \leq C\,\|\Omega\|_{L^2} \tag{5.22}$$

and then for any map v in $L^2(B^4, \mathbb{R}^m)$ the following equivalence holds

$$-\Delta v = \Omega\, v \Longleftrightarrow \mathrm{div}\,(A\nabla v - \nabla A\, v) = 0. \tag{5.23}$$

Having now the equation $-\Delta v = \Omega\, v$ in the form

$$\mathrm{div}\left(\nabla w - 2\nabla A\, A^{-1}\, w\right) = 0$$

where $w := A\,v$ permits to obtain easily the following Morrey estimate: $\forall\,\rho < 1$

$$\forall\,\rho < 1 \quad \sup_{x_0 \in B_\rho(0),\ r < 1-\rho} r^{-\nu} \int_{B_r(x_0)} |w|^2 < +\infty, \tag{5.24}$$

for some $\nu > 0$. As in the previous sections one deduces using Adams embeddings that $w \in L^q_{\mathrm{loc}}$ for some $q > 2$. Bootstrapping this information in the equation gives $v \in L^\infty_{\mathrm{loc}}$ (see [26] for a complete description of these arguments).

5.1 Concluding remarks

More jacobian structures or anti-symmetric structures have been discovered in other conformally invariant problems such as Willmore surfaces [24], bi-harmonic maps into manifolds [22], n/2-harmonic maps into manifolds [7–9], etc. Applying then integrability by compensation results in the spirit of what has been presented above, analysis questions such as the regularity of weak solutions, the behavior of sequences of solutions or the compactness of Palais-Smale sequences.... have been solved in these works.

Moreover, beyond the conformal dimension, while considering the same problems, but in dimension larger than the conformal one, similar approaches can be very efficient and the same strategy of proofs can sometimes be developed successfully (see for instance [27]).

References

[1] D. R. ADAMS, *A note on Riesz potentials*, Duke Math. J. **42** (1975), 765–778.

[2] F. BETHUEL, *Un résultat de régularité pour les solutions de l'équation de surfaces à courbure moyenne prescrite*, (French) "A regularity result for solutions to the equation of surfaces of prescribed mean curvature" C. R. Acad. Sci. Paris Sér. I Math. **314** (1992), 1003–1007.

[3] F. BETHUEL, *On the singular set of stationary harmonic maps*, Manuscripta Math. **78** (1993), 417–443.

[4] P. CHONÉ, *A regularity result for critical points of conformally invariant functionals*, Potential Anal. **4** (1995), 269–296.

[5] R. R. COIFMAN, P.-L. LIONS, Y. MEYER and S. SEMMES, *Compensated compactness and Hardy spaces*, J. Math. Pures Appl. (9) **72** (1993), 247–286.

[6] R. R. COIFMAN, R. ROCHBERG and G. WEISS, *Factorization theorems for Hardy spaces in several variables*, Ann. of Math. (2) **103** (1976), 611–635.

[7] F. DA LIO, "Fractional Harmonic Maps into Manifolds in odd Dimension $n > 1$", arXiv, 2010.

[8] F. DA LIO and T. RIVIÈRE, *3-Commutators estimates and the regularity of $1/2$ harmonic maps into spheres*, Analysis & PDE **4** (2011), 149–190.

[9] F. DA LIO and T. RIVIÈRE, *Sub-criticality of non-local Schrödinger systems with antisymmetric potentials and applications to half-harmonic maps*, Adv. Math. **227** (2011), 1300–1348.

[10] C. EVANS, *Partial regularity for stationary harmonic maps into spheres*, Arch. Rat. Mech. Anal. **116** (1991), 101–113.

[11] J. FREHSE, *A discontinuous solution of a midly nonlinear elliptic system*, Math. Z. **134** (1973), 229–230.

[12] A. FREIRE, S. MÜLLER and M. STRUWE, *Weak convergence of wave maps from $(1+2)$-dimensional Minkowski space to Riemannian manifolds*, Invent. Math. **130** (1997), 589–617.

[13] Y. GE, *Estimations of the best constant involving the L^2 norm in Wente's inequality and compact $H-$Surfaces in Euclidian space*, ESAIM: COCV **3** (1998), 263–300.

[14] M. GIAQUINTA, "Multiple Integrals in the Calculus of Variations and Nonlinear Elliptic Systems", Annals of Mathematics Studies, 105, Princeton University Press, Princeton, NJ, 1983.

[15] M. GRÜTER, *Conformally invariant variational integrals and the removability of isolated singularities*, Manuscripta Math. **47** (1984), 85–104.

[16] M. GRÜTER, *Regularity of weak H-surfaces*, J. Reine Angew. Math. **329** (1981), 1–15.

[17] E. HEINZ, *Ein Regularitätssatz für schwache Lösungen nichtlinearer elliptischer systeme*, (German) Nachr. Akad. Wiss. Göttingen Math.-Phys. Kl. **II** 1975, 1–13.

[18] E. HEINZ, *Über die regularität schwacher Lösungen nichtlinearer elliptischer systeme*, (German) "On the regularity of weak solutions of nonlinear elliptic systems" Nachr. Akad. Wiss. Göttingen Math.-Phys. Kl. **II** (1986), 1–15.

[19] F. HÉLEIN, "Harmonic Maps, Conservation Laws and Moving Frames", Cambridge Tracts in Math. 150, Cambridge Univerity Press, 2002.

[20] S. HILDEBRANDT, "Nonlinear Elliptic Systems and Harmonic Mappings", Proceedings of the 1980 Beijing Symposium on Differential Geometry and Differential Equations, vol. 1, 2, 3, Beijing, 1980, 481–615, Science Press, Beijing, 1982.

[21] S. HILDEBRANDT, "Quasilinear elliptic systems in diagonal form", Systems of nonlinear partial differential equations (Oxford, 1982), 173-217, NATO Adv. Sci. Inst. Ser. C Math. Phys. Sci., **111**, Reidel, Dordrecht, 1983.

[22] T. LAMM and T. RIVIÈRE, *Conservation laws for fourth order systems in four dimensions*, Comm. P.D.E. **33** (2008), 245–262.

[23] T. MÜLLER, *Higher integrability of determinants and weak convergence in L^1*, J. Reine Angew. Math. **412** (1990), 20–34.

[24] T. RIVIÈRE, *Conservation laws for conformally invariant variational problems*, Invent. Math. **168** (2007), 1–22.

[25] T. RIVIÈRE, *Analysis aspects of Willmore surfaces*, Inventiones Math. **174** (2008), 1–45.

[26] T. RIVIÈRE, *Sub-criticality of Schrödinger systems with antisymmetric potentials*, J. Math. Pures Appl. (9) **95** (2011), 260–276.

[27] T. RIVIÈRE and M. STRUWE, *Partial regularity for harmonic maps and related problems*, Comm. Pure Appl. Math. **61** (2008), 451–463.

[28] J. SHATAH, *Weak solutions and development of singularities of the SU(2) σ-model*, Comm. Pure Appl. Math. **41** (1988), 459–469.
[29] J. SHATAH and M. STRUWE, *The Cauchy problem for wave maps*, Int. Math. Res. Not. (2002), 555–571.
[30] T. TAO, *Global regularity of wave maps. I. Small critical Sobolev norm in high dimension*, Internat. Math. Res. Notices (2001), 299–328.
[31] T. TAO, *Global regularity of wave maps. II. Small energy in two dimensions*, Comm. Math. Phys. **224** (2001), 443–544.
[32] L. TARTAR, "Remarks on Oscillations and Stokes' Equation. Macroscopic Modelling of Turbulent Flows", Nice, 1984, 24–31, Lecture Notes in Phys., **230**, Springer, Berlin, 1985.
[33] L. TARTAR, "An Introduction to Sobolev Spaces and Interpolation Spaces", Lecture Notes of the Unione Matematica Italiana **3**, Springer, Berlin; UMI, Bologna, 2007.
[34] P. TOPPING, *The optimal constant in Wente's L^∞ estimate*, Comm. Math. Helv. **72** (1997), 316–328.
[35] K. K. UHLENBECK, *Connections with L^p bounds on curvature*, Comm. Math. Phys. **83** (1982), 31–42.
[36] H. C. WENTE, *An existence theorem for surfaces of constant mean curvature*, J. Math. Anal. Appl. **26** (1969), 318–344.

Regularity properties of equilibrium configurations of epitaxially strained elastic films

Bruno De Maria and Nicola Fusco

Abstract. We consider a variational model introduced in the physical literature to describe the epitaxial growth of an elastic film over a rigid substrate, when a lattice mismatch between the two materials is present. We establish the regularity of volume constrained local minimizers of the total energy, proving in particular the so called zero contact-angle condition between the film and the substrate.

Contents **169**

1 Introduction

In this paper we present some regularity results for equilibrium configurations of a variational energy modeling the epitaxial deposition of a film onto a rigid substrate in presence of a mismatch strain between the lattice parameters of the two crystalline solids.

The presence of such a strain is responsible of the so called Asaro-Grinfeld-Tiller instability of the flat configuration: to release some of the elastic energy due to the strain the atoms on the free surface of the film tend to rearrange into a more favorable configuration. Thus, the film profile becomes wavy or breaks into several material clusters, also known as *islands*, separated by a thin layer that wets the substrate. However, this phenomenon occurs only when the thickness of the film reaches a critical threshold. We refer to [2, 8] for a detailed account on this effect.

Among the several atomistic and continuum models available for the growth of epitaxially strained thin films, we follow here a variational approach contained in [14] (see also [12, 15]), which was first studied from an analytical point of view by Bonnetier and Chambolle in [3]. As in this paper we restrict to two-dimensional morphologies, corresponding to three-dimensional configurations with planar symmetry.

To describe the model studied in [3] we start by introducing the reference configuration of the film

$$\Omega_h := \left\{z = (x, y) \in \mathbb{R}^2 : 0 < x < b,\ 0 < y < h(x)\right\},$$

where $h : [0, b] \to [0, \infty)$ represents the free-profile of the film. Following the physical literature, the function h will be assumed to be b-periodic, *i.e.*, $h(0) = h(b)$. Denoting by $u : \Omega_h \to \mathbb{R}^2$ the planar displacement of the film, we shall assume that $u(x, 0) = e_0(x, 0)$ at the interface between the film and the substrate, where the constant e_0 depends on the gap between the lattices of the two materials. In view of the periodicity assumption on h, we shall also impose the periodicity condition $u(b, y) = u(0, y) + e_0(b, 0)$.

For a smooth configuration (h, u) the energy considered in [3] is given by

$$G(h, u) = \int_{\Omega_h} \left[\mu |E(u)|^2 + \frac{\lambda}{2}(\mathrm{div} u)^2\right] dz + \sigma \mathcal{H}^1(\Gamma_h), \tag{1.1}$$

where μ and λ represent the Lamé coefficients of the material, $E(u) := (\nabla u + \nabla^T u)/2$ is the linearized elastic strain, σ is the surface tension on the profile of the film, Γ_h is the graph of h, and $\mathcal{H}^1$ denotes the one-dimensional Hausdorff measure.

The equilibrium configurations are defined as the minimizers of the above energy under a volume constraint $|\Omega_h| = d$. However, as shown in [3], the minimizing sequences of the functional G may converge to more irregular profiles h, which are only lower semicontinuous functions of bounded variation. Denoting by $AP(0, b)$ (see (2.1)) the class of such admissible profiles, a standard relaxation procedure leads to the following representation of the energy associated to a configuration (h, u), where $h \in AP(0, b)$, $E(u) \in L^2(\Omega; \mathbb{M}^{2\times 2}_{\mathrm{sym}})$, and u satisfies the Dirichlet and periodicity conditions stated above,

$$F(h, u) = \int_{\Omega_h} \left[\mu |E(u)|^2 + \frac{\lambda}{2}(\mathrm{div} u)^2\right] dz + \sigma \mathcal{H}^1(\Gamma_h) + 2\sigma \mathcal{H}^1(\Sigma_h).$$

Here Γ_h denotes the extended graph of h and Σ_h is the union of all vertical cuts, whose length is counted twice, since they arise as limits of regular profiles. The existence of minimizers for F under the volume constraint $|\Omega_h| = d$ follows immediately as a result of its definition via relaxation.

In this paper we address the regularity of global and local constrained minimizers of F. Our approach follows very closely the one introduced by Fonseca, Leoni, Morini and the second author in [6] (see also [5] for a related problem), where a slightly different model and a stronger notion of local minimality were considered. As in [6] we show that the profile of a constrained local minimizer of F may have at most finitely many cut segments or singularities of cusp type, being C^1-regular away from these singular points (see Theorem 2.5 below). In particular, we show that the so called *zero contact-angle condition* also holds for the model studied here.

The notion of local minimality considered in this paper was introduced in [7] in connection with a thorough study of the local and global minimality properties of the flat configuration. As we have already mentioned, this notion is weaker than the one considered in [6]. As a consequence, the starting points of the regularity analysis, that is the equivalence between local minimality for the constrained problem and the penalized one and the interior ball condition, are more delicate to prove (see Proposition 3.1 and Theorem 3.4). Also the proof of the C^1-regularity at the interface between the film and the substrate presents additional technical difficulties because of the coupling of the Dirichlet condition on $\{y = 0\}$ and the Neumann condition on Γ_h that must be satisfied by the displacement u.

This paper grew out from a series of lectures given by the second author during the intensive period "Regularity for Non-linear PDEs", held in 2009 at the De Giorgi Center in Pisa. Both authors gratefully acknowledge the hospitality of the organizers and of the Center.

2 Statement of the existence and regularity results

In this section we present the model studied by Bonnetier and Chambolle in [3] and the related functional setting. We also state the regularity theorem.

Throughout the paper we denote by $z = (x, y)$ the generic point in $\mathbb{R}^2$ and by $B_r(z)$ the open disc centered in z with radius r. Given two sets $A, B \subset \mathbb{R}^2$, their *Hausdorff distance* is defined as

$$d_{\mathcal{H}}(A, B) := \inf\{\varepsilon > 0 : A \subset \mathcal{N}_\varepsilon(B) \text{ and } B \subset \mathcal{N}_\varepsilon(A)\},$$

where $\mathcal{N}_\varepsilon(A)$ denotes the ε-neighborhood of A.

In the following we are going to consider lower semicontinuous periodic profiles. To this aim we introduce the class of admissible profiles

$$AP(0, b) := \big\{g : \mathbb{R} \to [0, +\infty) : g \text{ is lower semicontinuous and } b\text{-periodic, } \operatorname{Var}(g; 0, b) < \infty\big\}, \tag{2.1}$$

where $\mathrm{Var}(g; 0, b)$ denotes the *pointwise total variation* of g over the interval $(0, b)$

$$\mathrm{Var}(g; 0, b) = \sup\left\{\sum_{i=1}^{n} |g(x_i) - g(x_{i-1})| : 0 \le x_0 < x_1 < \dots x_n \le b\right\}.$$

Since $g \in AP(0, b)$ is b-periodic, its pointwise total variation is finite over any bounded interval of $\mathbb{R}$. Therefore, it admits right and left limits at every $x \in \mathbb{R}$ denoted by $g(x+)$ and $g(x-)$, respectively. In the following we use the notation

$$g^+(x) := \max\{g(x+), g(x-)\}, \quad g^-(x) := \min\{g(x+), g(x-)\}. \quad (2.2)$$

To represent the region occupied by the film, we set

$$\Omega_g := \{(x, y) : x \in (0, b),\ 0 < y < g(x)\},$$
$$\Omega_g^\# := \{(x, y) : x \in \mathbb{R},\ 0 < y < g(x)\},$$

while the profile of the film is given by

$$\Gamma_g := \{(x, y) : x \in [0, b),\ g^-(x) \le y \le g^+(x)\}.$$

The set of vertical cracks (*or cuts*) is

$$\Sigma_g := \{(x, y) : x \in [0, b),\ g(x) < g^-(x),\ g(x) \le y < g^-(x)\}.$$

We will also use the notation $\Gamma_g^\# := \{(x, y) \in \mathbb{R}^2 : x \in \mathbb{R},\ g^-(x) \le y \le g^+(x)\}$. The set $\Sigma_g^\#$ is defined similarly. Finally, we set $\widetilde{\Gamma}_g = \Gamma_g \cup \Sigma_g$, $\widetilde{\Gamma}_g^\# = \Gamma_g^\# \cup \Sigma_g^\#$.

Given $g \in AP(0, b)$, we denote

$$LD_\#(\Omega_g; \mathbb{R}^2) := \big\{v \in L^2_{\mathrm{loc}}(\Omega_g^\#; \mathbb{R}^2) : v(x, y) = v(x+b, y) \text{ for } (x, y) \in \Omega_g^\#,\ E(v)|_{\Omega_g} \in L^2(\Omega_g; \mathbb{M}^{2\times 2}_{\mathrm{sym}})\big\},$$

where $E(v) := \frac{1}{2}(\nabla v + \nabla^T v)$, ∇v being the distributional gradient of v and $\nabla^T v$ its transpose. Given $e_0 \neq 0$, we define

$$X(e_0; b) := \big\{(g, v) : g \in AP(0, b),\ v : \Omega_g^\# \to \mathbb{R}^2 \text{ such that } v(\cdot, \cdot) - e_0(\cdot, 0) \in LD_\#(\Omega_g; \mathbb{R}^2),\ v(x, 0) = (e_0 x, 0) \text{ for all } x \in \mathbb{R}\big\}.$$

We introduce the following convergence in $X(e_0; b)$.

Definition 2.1. We say that $(h_n, u_n) \to (h, u)$ in $X(e_0; b)$ if the following two conditions hold:

(i) $\sup_n \mathrm{Var}(h_n; 0, b) < +\infty$ and $d_{\mathcal{H}}(\mathbb{R}^2_+ \setminus \Omega^\#_{h_n}, \mathbb{R}^2_+ \setminus \Omega^\#_h) \to 0$, where $\mathbb{R}^2_+ := \{(x, y) \in \mathbb{R}^2 : y \geq 0\}$;

(ii) $u_n \rightharpoonup u$ weakly in $W^{1,2}_{\mathrm{loc}}(\Omega^\#_h; \mathbb{R}^2)$.

Notice that the definition is well posed since by (i) it follows that if $\mathrm{dist}(\Omega', \mathbb{R}^2_+ \setminus \Omega^\#_h) > 0$ then $\mathrm{dist}(\Omega', \mathbb{R}^2_+ \setminus \Omega^\#_{h_n}) > 0$ for n large enough. The notion of convergence just introduced is motivated by the following compactness result (see [3, 6]).

Theorem 2.2. *Let $(h_n, u_n) \in X(e_0; b)$ be such that*

$$\sup_n \left\{ \int_{\Omega_{h_n}} |E(u_n)|^2 + \mathrm{Var}(h_n; 0, b) + |\Omega_{h_n}| \right\} < \infty.$$

Then there exist $(h, u) \in X(e_0; b)$ and a subsequence (h_{n_k}, u_{n_k}) such that $(h_{n_k}, u_{n_k}) \to (h, u)$ in $X(e_0; b)$.

We work in the framework of linearized elasticity and we consider isotropic and homogeneous materials. Hence, the elastic energy density $W : \mathbb{M}^{2\times 2}_{\mathrm{sym}} \to [0, +\infty)$ takes the form

$$W(\xi) := \frac{1}{2}\mathbb{C}\xi : \xi = \mu|\xi|^2 + \frac{\lambda}{2}\big[\mathrm{tr}(\xi)\big]^2,$$

where

$$\mathbb{C}\xi = \begin{pmatrix} (2\mu + \lambda)\xi_{11} + \lambda\xi_{22} & 2\mu\xi_{12} \\ 2\mu\xi_{12} & (2\mu + \lambda)\xi_{22} + \lambda\xi_{11} \end{pmatrix}$$

and the *Lamé coefficients* μ and λ satisfy the ellipticity conditions

$$\mu > 0 \qquad \text{and} \qquad \lambda > -\mu. \tag{2.3}$$

Note that

$$W(\xi) \geq \min\{\mu, \mu + \lambda\}|\xi|^2 \qquad \text{for all } \xi \in \mathbb{M}^{2\times 2}_{\mathrm{sym}} \tag{2.4}$$

and thus condition (2.3) guarantees that W is coercive. If $(g, v) \in X(e_0; b)$ and g is Lipschitz, the total free energy is defined as

$$G(g, v) := \int_{\Omega_g} W(E(v))\,dz + \mathcal{H}^1(\Gamma_g),$$

where we have set $\sigma = 1$ in (1.1). The following result, proved in [3] (see also [6]), gives a representation formula for the energy in the general case and an existence result for the corresponding constrained minimum problem. To this aim, we set for any $(g, v) \in X(e_0; b)$

$$F(g, v) := \inf\Big\{\liminf_{n\to\infty} G(g_n, v_n) : (g_n, v_n) \to (g, v) \text{ in } X(e_0; b),\\ g_n \text{ Lipschitz}, |\Omega_{g_n}| = |\Omega_g|\Big\}.$$

Theorem 2.3. *For any pair* $(g, v) \in X(e_0; b)$

$$F(g, v) = \int_{\Omega_g} W(E(v))\,dz + \mathcal{H}^1(\Gamma_g) + 2\mathcal{H}^1(\Sigma_g).$$

Moreover, for any $d > 0$ *the minimum problem*

$$\min\{F(g, v) : (g, v) \in X(e_0; b),\ |\Omega_g| = d\} \tag{2.5}$$

has a solution, the minimum value in (2.5) *is equal to*

$$\inf\{G(g, v) : (g, v) \in X(e_0; b),\ |\Omega_g| = d,\ g \textit{ Lipschitz}\}$$

and the limit points of minimizing sequences are minimizers of (2.5).

Our regularity result applies not only to *global minimizer, i.e.* minimizers of (2.5), but also to *local minimizers*, which are defined as follows.

Definition 2.4. We say that an admissible pair $(h, u) \in X(e_0; b)$ is a *local minimizer* for F if there exists $\delta > 0$ such that

$$F(h, u) \leq F(g, v)$$

for all pairs $(g, v) \in X(e_0; b)$, with $|\Omega_g| = |\Omega_h|$ and $d_{\mathcal{H}}(\widetilde{\Gamma}^{\#}_h, \widetilde{\Gamma}^{\#}_g) < \delta$.

Notice that a (sufficiently regular) local minimizer $(h, u) \in X(e_0; b)$ satisfies the following set of Euler-Lagrange conditions:

$$\begin{cases} \operatorname{div} \mathbb{C}E(u) = 0 & \text{in } \Omega_h; \\ \mathbb{C}E(u)[\nu] = 0 & \text{on } \Gamma_h \cap \{y > 0\}; \\ \mathbb{C}E(u)(0, y)[\nu] = -\mathbb{C}E(u)(b, y)[\nu] & \text{for } 0 < y < h(0) = h(b); \\ k + W(E(u)) = \text{const} & \text{on } \Gamma_h \cap \{y > 0\}, \end{cases} \tag{2.6}$$

where ν denotes the outer unit normal to Ω_h and k is the curvature of Γ_h. Due to (2.4), equation $(2.6)_1$ is a linear elliptic system satisfying the Legendre-Hadamard condition.

Before stating the regularity result, we need to introduce the set of *cusp points* of a function $g \in AP(0, b)$

$$\Sigma_{g,c} := \{(x, g(x)) : x \in [0, b), g^-(x) = g(x), \text{ and } g'_+(x) = -g'_-(x) = +\infty\},$$

where g^- is defined in (2.2), and g'_+ and g'_- denote the right and left derivatives, respectively. As before, the set $\Sigma^\#_{g,c}$ is obtained by replacing $[0, b)$ by $\mathbb{R}$ in the previous formula and coincides with the b-periodic extension of $\Sigma_{g,c}$.

Theorem 2.5 (Regularity of local minimizers). *Let $(h, u) \in X(e_0; b)$ be a local minimizer for F. Then the following regularity results hold:*

(i) *there are at most finitely many cusp points and vertical cracks in $[0, b)$;*
(ii) *the curve $\Gamma^\#_h$ is of class C^1 away from $\Sigma^\#_h \cup \Sigma^\#_{h,c}$;*
(iii) *$\Gamma^\#_h \cap \{(x, y) : y > 0\}$ is of class $C^{1,\alpha}$ away from $\Sigma^\#_h \cup \Sigma^\#_{h,c}$ for all $\alpha \in (0, 1/2)$;*
(iv) *let $A := \{x \in \mathbb{R} : h(x) > 0$ and h is continuous at $x\}$. Then A is an open set of full measure in $\{h > 0\}$ and h is analytic in A.*

Statement (ii) of Theorem 2.5 implies in particular that the zero contact-angle condition between film and substrate holds. On the other hand, if $h > 0$ is of class $C^{1,\alpha}$ for all $\alpha \in (0, 1/2)$, and if $(h, u) \in X(e_0; b)$ satisfies the first three equations in (2.6), then classical elliptic regularity results (see [7, Proposition 8.9]) imply that $u \in C^{1,\alpha}(\overline{\Omega}_h)$ for all $\alpha \in (0, 1/2)$. Moreover, if also $(2.6)_4$ holds in the distributional sense, then the results contained in [11, Subsection 4.2] imply that (h, u) is analytic.

3 The interior ball condition

The proof of Theorem 2.5 is quite long and we shall follow the path set in [6]. When dealing with the parts of $\Gamma^\#_h$ above the x-axis, our argument is a bit simpler than the one followed in [6]. On the other hand, the proof of the zero contact-angle condition for the model considered here requires some new ideas and extra care.

The first step consists in removing the constraint $|\Omega_h| = d$ by showing that if (h, u) is a local minimizer, then (h, u) is also a local minimizer of the penalized functional

$$(g, v) \in X(e_0; b) \mapsto F(g, v) + \Lambda \big| |\Omega_g| - d \big|,$$

for some $\Lambda > 0$ sufficiently large, depending also on u. This gives a much larger choice of variations and in particular allows us to prove that

$\Omega_h^\#$ satisfies a uniform interior ball condition, namely that if $0 < \varrho < 1/\Lambda$ is sufficiently small (depending on u), then for all $z \in \widetilde{\Gamma}_h^\#$ there exists an open disk $B_\varrho(z_0) \subset \Omega_h^\#$ such that $\partial B_\varrho(z_0) \cap \widetilde{\Gamma}_h^\# = \{z_0\}$.

Proposition 3.1. *Let $(h, u) \in X(e_0; b)$ be a local minimizer for F. Then, there exist $\sigma_0, \Lambda_0 > 0$ such that*

$$F(h,u) = \min\big\{F(g,v) + \Lambda\big|d - |\Omega_g|\big| : (v,g) \in X(e_0;b),\; d_{\mathcal{H}}(\widetilde{\Gamma}_g, \widetilde{\Gamma}_h) \le \sigma\big\}, \tag{3.1}$$

for all $0 < \sigma \le \sigma_0$, $\Lambda \ge \Lambda_0$.

Proof. Observe that from Theorem 2.2 the existence of a minimizer (g, v) for the variational problem in (3.1) follows immediately. Moreover, since $F(g,v) + \Lambda\big|d - |\Omega_g|\big| \le F(h,u)$, we have

$$\big|d - |\Omega_g|\big| \le \frac{F(h,u)}{\Lambda}, \qquad \text{and} \qquad \mathcal{H}^1(\widetilde{\Gamma}_g) \le F(h,u). \tag{3.2}$$

Next, given any function $k \in AP(0,b)$ and any positive number M, let us denote by $k^M = \min\{k, M\}$ the function obtained by truncating k at the level M. Clearly, $k^M \in AP(0,b)$. Moreover, it can be easily checked that if $k, l \in AP(0,b)$ and $M > 0$, then

$$d_{\mathcal{H}}(\widetilde{\Gamma}^\#_{k^M}, \widetilde{\Gamma}^\#_{l^M}) \le d_{\mathcal{H}}(\widetilde{\Gamma}^\#_k, \widetilde{\Gamma}^\#_l). \tag{3.3}$$

Let $\delta > 0$ be as in Definition 2.4. We now choose $\sigma_1 > 0$ with the property that

$$\|h - h^M\|_{L^1(0,b)} < 3\sigma_1 \implies d_{\mathcal{H}}(\widetilde{\Gamma}^\#_{h^M}, \widetilde{\Gamma}^\#_h) < \frac{\delta}{2}. \tag{3.4}$$

Next, we claim that there exists $\sigma_0 \in (0, \delta/2)$ such that if $g \in AP(0,b)$, then

$$\mathcal{H}^1(\widetilde{\Gamma}^\#_g) \le F(h,u), \quad \big|d - |\Omega_g|\big| \le F(h,u), \quad d_{\mathcal{H}}(\widetilde{\Gamma}^\#_g, \widetilde{\Gamma}^\#_h) \le \sigma_0 \Longrightarrow \|h - g\|_{L^1(0,b)} < \sigma_1. \tag{3.5}$$

To prove the claim we argue by contradiction, assuming that there exists a sequence $g_n \in AP(0,b)$ such that $\mathcal{H}^1(\widetilde{\Gamma}_{g_n}) \le F(h,u)$, $\big|d - |\Omega_{g_n}|\big| \le F(h,u)$ for all n, $d_{\mathcal{H}}(\widetilde{\Gamma}^\#_{g_n}, \widetilde{\Gamma}^\#_h) \to 0$ as $n \to \infty$, but $\|h - g_n\|_{L^1(0,b)} \ge \sigma_1$. Thus, we may assume that, up to a subsequence, g_n converge in $L^1(0,b)$ and a.e. to a function $g \in AP(0,b)$. Note that , since $d_{\mathcal{H}}(\widetilde{\Gamma}^\#_{g_n}, \widetilde{\Gamma}^\#_h) \to 0$, then for all $x \in (0,b)$ such that $g_n(x) \to g(x)$, we have $(x, g(x)) \in \widetilde{\Gamma}_h$.

Therefore $g = h$ a.e. in $(0, b)$, thus contradicting the fact that $\|h - g_n\|_{L^1(0,b)} \geq \sigma_1$ for all n.

To prove the assertion it is enough to show that there exists $\Lambda_0 \geq 1$ such that for any $\sigma \in (0, \sigma_0]$ and any $\Lambda \geq \Lambda_0$, the minimizer (g, v) of the minimum problem in (3.1) satisfies the volume constraint $|\Omega_g| = d$. To this aim, we consider two cases.

If $|\Omega_g| > d$, we set $\tilde{g} = g^M$, where M is such that $|\Omega_{g^M}| = d$, and $\tilde{v} = v$. We have

$$F(\tilde{g}, \tilde{v}) < F(g, v) + \Lambda\big|d - |\Omega_g|\big| \leq F(h, u). \tag{3.6}$$

Moreover, from (3.2) and (3.5) it follows that $\|h - g\|_{L^1(0,b)} < \sigma_1$. Thus, from (3.2) we get

$$\begin{aligned}
\|h - h^M\|_{L^1(0,b)} &\leq \|h - g\|_{L^1(0,b)} + \|g - g^M\|_{L^1(0,b)} + \|g^M - h^M\|_{L^1(0,b)} \\
&\leq 2\|h - g\|_{L^1(0,b)} + \|g - g^M\|_{L^1(0,b)} \\
&\leq 2\sigma_1 + \big|d - |\Omega_g|\big| \leq 2\sigma_1 + \frac{F(h, u)}{\Lambda} < 3\sigma_1,
\end{aligned}$$

provided that we choose $\Lambda_0 > F(h, u)/\sigma_1$. Therefore, with such a choice of Λ_0, recalling (3.4) and that $\sigma_0 < \delta/2$ we have that

$$d_{\mathcal{H}}(\widetilde{\Gamma}^{\#}_{\tilde{g}}, \widetilde{\Gamma}^{\#}_{h}) \leq d_{\mathcal{H}}(\widetilde{\Gamma}^{\#}_{g^M}, \widetilde{\Gamma}^{\#}_{h^M}) + d_{\mathcal{H}}(\widetilde{\Gamma}^{\#}_{h}, \widetilde{\Gamma}^{\#}_{h^M}) \leq d_{\mathcal{H}}(\widetilde{\Gamma}^{\#}_{g}, \widetilde{\Gamma}^{\#}_{h}) + \frac{\delta}{2} < \delta,$$

where in the second inequality we have used (3.3). Hence, recalling (3.6), we get a contradiction to the minimality of (h, u). This proves that $|\Omega_g| \leq d$.

Let us now show that also inequality $|\Omega_g| < d$ cannot hold if Λ_0 is large enough. In fact, if $|\Omega_g| < d$ we may define $\widetilde{g} = g + \frac{d - |\Omega_g|}{b}$. Note that with such a choice $|\Omega_{\tilde{g}}| = d$. Moreover, using (3.2) again,

$$d_{\mathcal{H}}(\widetilde{\Gamma}^{\#}_{\tilde{g}}, \widetilde{\Gamma}^{\#}_{h}) \leq d_{\mathcal{H}}(\widetilde{\Gamma}^{\#}_{\tilde{g}}, \widetilde{\Gamma}^{\#}_{g}) + d_{\mathcal{H}}(\widetilde{\Gamma}^{\#}_{g}, \widetilde{\Gamma}^{\#}_{h}) < \frac{d - |\Omega_g|}{b} + \frac{\delta}{2} < \delta, \tag{3.7}$$

if we choose $\Lambda_0 > 2F(h, u)/(b\delta)$. Thus if we set

$$\tilde{v}(x, y) = \begin{cases} e_0(x, 0) & \text{if } 0 < y < \dfrac{d - |\Omega_g|}{b} \\[2ex] v\left(x, y - \dfrac{d - |\Omega_g|}{b}\right) & \text{if } y > \dfrac{d - |\Omega_g|}{b}, \end{cases}$$

we get, using the minimality of (g, v),

$$\begin{aligned}
F(\tilde{g}, \tilde{v}) - F(h, u) &= F(g, v) + \frac{e_0^2}{2}(2\mu + \lambda)(d - |\Omega_g|) - F(h, u) \\
&\leq \frac{e_0^2}{2}(2\mu + \lambda)(d - |\Omega_g|) - \Lambda(d - |\Omega_g|) < 0,
\end{aligned}$$

provided we choose $\Lambda_0 > e_0^2(2\mu + \lambda)/2$. This inequality, together with (3.7), contradicts the local minimality of (h, u), thus proving the assertion. □

Before going on with the proof of the interior ball condition we state a simple approximation result for one-dimensional BV-functions.

Lemma 3.2. *Let $h : [0, b] \to \mathbb{R}$ be a function with finite total variation. There exists a sequence of Lipschitz functions $g_n : [0, b] \to \mathbb{R}$ such that $g_n(0) = h(0)$, $g_n(b) = h(b)$, $g_n \to h$ in $L^1(0, b)$ and*

$$\mathcal{H}^1(\Gamma_{g_n}) \to \mathcal{H}^1(\Gamma_h \cap ((0, b) \times \mathbb{R})) + |h(0+) - h(0)| + |h(b-) - h(b)|.$$

Proof. Given $\varepsilon > 0$, extend h to a function $h_\varepsilon : (-\varepsilon, b + \varepsilon) \to \mathbb{R}$, by setting $h_\varepsilon(x) := h(0)$ for $x < 0$ and $h_\varepsilon(x) := h(b)$ if $x > b$. By a well-known property of BV functions, there exists $g_\varepsilon : (-\varepsilon, b + \varepsilon) \to \mathbb{R}$, Lipschitz continuous, such that

$$\|h_\varepsilon - g_\varepsilon\|_{L^1(-\varepsilon,b+\varepsilon)} < \varepsilon \quad \text{and} \quad |\mathcal{H}^1(\Gamma_{g_\varepsilon}) - H^1(\Gamma_{h_\varepsilon})| < \varepsilon,$$

where $\Gamma_{h_\varepsilon} := \{(x, y) : x \in (-\varepsilon, b + \varepsilon), h_\varepsilon^-(x) \le y \le h_\varepsilon^+(x)\}$ and Γ_{g_ε} is the graph of g_ε. Moreover, by slightly modifying g_ε in a neighborhood of 0 and of b, if necessary, we may also assume that $g_\varepsilon(0) = h(0)$ and that $g_\varepsilon(b) = h(b)$. Then, $g_{\varepsilon|[0,b]}$ gives the required approximation of h in $[0, b]$. □

Next result contains a useful isoperimetric inequality, whose elementary proof is reproduced from [7, Lemma 6.6].

Lemma 3.3. *Let $k \in AP(0,b)$ and let $B_\rho(z_0)$ be a disk such that $B_\rho(z_0) \subset (x, y) : x \in \mathbb{R}$ and $y < k(x)\}$ and $\partial B_\rho(z_0) \cap \widetilde{\Gamma}_k^\#$ contains $z_1 = (x_1, y_1)$, $z_2 = (x_2, y_2)$. Let γ be the shortest arc on $\partial B_\rho(z_0)$ connecting z_1 and z_2 (any of the two possible arcs if z_1 and z_2 are antipodal) and let γ' be the arc on $\widetilde{\Gamma}_k^\#$ connecting z_1 and z_2. Then*

$$\mathcal{H}^1(\gamma') - \mathcal{H}^1(\gamma) \ge \frac{1}{\rho}|D|,$$

where D is the region enclosed by $\gamma \cup \gamma'$.

Proof. Denote by h the function whose graph coincides with γ and observe that $\frac{1}{\rho} = -\left(\frac{h'}{\sqrt{1+h'^2}}\right)'$. If $k : [x_1, x_2] \to \mathbb{R}$ is any Lipschitz function such that $k(x_1) = y_1$ and $k(x_2) = y_2$, we have

$$\mathcal{H}^1(\gamma') - \mathcal{H}^1(\gamma) = \int_{x_1}^{x_2} \left(\sqrt{1+k'^2} - \sqrt{1+h'^2}\right) dx$$
$$\geq \int_{x_1}^{x_2} \frac{(k'-h')h'}{\sqrt{1+h'^2}}\, dx = \frac{1}{\rho}\int_{x_1}^{x_2} (k-h)\, dx.$$

Assume now that $k \in AP(0,b)$. Without loss of generality, we may also assume that $k(x_i) = y_i$ for $i = 1, 2$. Then, by approximating k in $[x_1, x_2]$ with a sequence k_n of Lipschitz functions according to Lemma 3.2 and passing to the limit in the above formula, we get

$$\mathcal{H}^1(\gamma') - \mathcal{H}^1(\gamma) \geq \frac{1}{\rho}\int_{x_1}^{x_2} (k-h)\, dx.$$

From this inequality, the result immediately follows by the assumption $k \geq h$. □

We are now in position to give the proof of the interior ball condition. Note that though the proof presented here is similar to the one given in [7, Lemma 6.7], the argument is more delicate since we are dealing with a more general notion of local minimizer.

Theorem 3.4. *Let $(h,u) \in X(e_0; b)$ be a b-periodic local minimizers for F. Then there exists $\rho_0 > 0$ such that for any $z \in \widetilde{\Gamma}_h^\#$ there exists a disk $B_{\rho_0}(z_0) \subset \{(x,y) : x \in \mathbb{R}$ and $y < h(x)\}$ with $\partial B_{\rho_0}(z_0) \cap \widetilde{\Gamma}_h^\# = \{z\}$.*

Proof.
Step 1. Let us first prove that if ρ_0 is sufficiently small then the boundary of every disk contained in $\{(x,y) : x \in \mathbb{R}$ and $y < h(x)\}$ can intersect $\widetilde{\Gamma}_h^\#$ at most in one point.

Let σ_0, Λ_0 be as in Proposition 3.1 and assume that there exists $B_\rho(z_0) \subset \{(x,y) : x \in \mathbb{R}$ and $y < h(x)\}$ such that $\partial B_\rho(z_0) \cap \widetilde{\Gamma}_h^\#$ contains two distinct points z_1, z_2. Setting $z_i = (x_i, y_i)$, $i = 0, 1, 2$, we may assume that $x_1 < x_0 < x_2$. Let $\bar{z}$ be a point in $\widetilde{\Gamma}_h^\# \cap ([x_1, x_2] \times \mathbb{R})$ maximizing the distance from $\partial B_\rho(z_0)$. We claim that if

$$\rho < \frac{\sigma_0}{2(1+\Lambda_0\sigma_0)} \tag{3.8}$$

then $\operatorname{dist}(\bar{z}, \partial B_\rho(z_0)) \leq \sigma_0$.

In fact, if the claim were not true, setting $\bar{z} = (\bar{x}, \bar{y})$, the line $\{y = \bar{y} - \sigma_0/2\}$ would lie above the disk $B_\rho(z_0)$. Moreover, since $\widetilde{\Gamma}_h^\# \cap ([x_1, x_2] \times \mathbb{R})$ is a connected arc, this line would intersect it in at least two points. Let us denote by $\bar{z}_1 = (\bar{x}_1, \bar{y} - \sigma_0/2)$ and $\bar{z}_2 = (\bar{x}_2, \bar{y} - \sigma_0/2)$ the points in $\widetilde{\Gamma}_h^\# \cap ([x_1, x_2] \times \mathbb{R}) \cap \{y = \bar{y} - \sigma_0/2\}$, with the smallest and largest abscissa, respectively. Note that the projection over the y-axis of the subarc of $\widetilde{\Gamma}_h^\#$ connecting $\bar{z}_1$ to $\bar{z}$ is longer than $\sigma_0/2$ and the same is true for the projection of the arc connecting $\bar{z}_2$ to $\bar{z}$. Therefore we have that

$$\mathcal{H}^1\big(\widetilde{\Gamma}_h^\# \cap \big([\bar{x}_1, \bar{x}_2] \times (\bar{y} - \sigma_0/2, +\infty)\big)\big) \geq \sigma_0, \tag{3.9}$$

while $\bar{x}_2 - \bar{x}_1 \leq 2\rho$. Let us now define

$$\tilde{h}(x) = \begin{cases} \min\left\{h(x), \bar{y} - \dfrac{\sigma_0}{2}\right\} & \text{if } x \in [\bar{x}_1, \bar{x}_2] \\ h(x) & \text{otherwise,} \end{cases}$$

and set $\tilde{u} = u$. Then, we have that $d_{\mathcal{H}}(\widetilde{\Gamma}_{\tilde{h}}^\#, \widetilde{\Gamma}_h^\#) \leq \sigma_0$. Moreover, from (3.9) we easily get, recalling also (3.8),

$$F(\tilde{h}, \tilde{u}) + \Lambda_0\big|d - |\Omega_{\tilde{h}}|\big| - F(h, u) \leq 2\rho + \rho\Lambda_0\sigma_0 - \sigma_0 < 0,$$

thus contradicting the fact that h minimizes the variational problem in (3.1).

Thus we have proved that if ρ satisfies (3.8) then every point in $\widetilde{\Gamma}_h^\# \cap ([x_1, x_2] \times \mathbb{R})$ lies at a distance from $\partial B_\rho(z_0)$ smaller than or equal to σ_0. Therefore, if we define $\tilde{h}$ to be the function obtained by replacing h in the interval $[x_1, x_2]$ with the function whose graph is the upper arc on $\partial B_\rho(z_0)$ connecting z_1 and z_2, we still have $d_{\mathcal{H}}(\widetilde{\Gamma}_{\tilde{h}}^\#, \widetilde{\Gamma}_h^\#) \leq \sigma_0$. Defining $\tilde{u}$ as above, we then have, using Lemma 3.3,

$$\begin{aligned} F(\tilde{h}, \tilde{u}) + \Lambda_0\big|d-|\Omega_{\tilde{h}}|\big| - F(h, u) &\\ \leq \mathcal{H}^1\big(\partial B_\rho(z_0) \cap ([x_1, x_2] \times \mathbb{R})\big) &\\ - \mathcal{H}^1\big(\widetilde{\Gamma}_h^\# \cap ([x_1, x_2] \times \mathbb{R})\big) + \Lambda_0|D| &\\ \leq -\frac{1}{\rho}|D| + \Lambda_0|D|, & \end{aligned}$$

a quantity which is negative if we impose that ρ is also smaller than $1/\Lambda_0$. In this case we get again a contradiction to the minimality of (h, u).

Step 2. We argue here as in the proof of [6, Proposition 3.3], whose argument goes as follows.

Fix $\rho_0 > 0$ so that the conclusion of Step 1 applies and consider the union U of all balls of radius ρ_0 that are contained in $\Omega := \{(x, y) : x \in \mathbb{R} \text{ and } y < h(x)\}$. Our thesis is equivalent to showing that $\Omega \subset U$. Assume by contradiction that this inclusion doesn't hold. Then there exist $z_0' \in \Omega \cap \partial U$, a sequence of balls $B_{\rho_0}(z_n) \subset \Omega$, and $z_n' \in \partial B_{\rho_0}(z_n)$ such that $z_n' \to z_0'$. Up to extracting a subsequence we may assume that $B_{\rho_0}(z_n) \to B_{\rho_0}(z_0)$ in the Hausdorff distance, for some ball $B_{\rho_0}(z_0) \subset \Omega$ having z_0' at its boundary. Note that the intersection of $\partial B_{\rho_0}(z_0)$ with $\widetilde{\Gamma}_h^\#$ must be nonempty, since otherwise we could slightly translate the ball, still remaining in Ω and this would violate the fact that $z_0' \in \partial U$. Hence, by the previous step, $\partial B_{\rho_0}(z_0) \cap \widetilde{\Gamma}_h^\# = \{z'\}$. If z_0' and z' are antipodal, then we can find $\tau > 0$ such that $B_{\rho_0}(z_0 + \tau(z_0' - z')) \subset \Omega$, which would imply that $z_0' \in U$, a contradiction. If z_0' and z' are not antipodal, then we can rotate $B_{\rho_0}(z_0)$ around z_0', slightly away from z', to get a ball B' of radius ρ_0 such that $\overline{B}' \subset \Omega$ and $z_0' \in \partial B'$. Translating now B' towards z_0' we find a ball of the same radius containing z_0' and contained in Ω, which gives again $z_0' \in U$. This concludes the proof of the theorem. □

The interior ball condition gives a sort of 'unilateral bound' on the curvature of $\widetilde{\Gamma}_h^\#$. This is a key point in proving Theorem 2.5. More precisely we have the following result whose elementary, but delicate proof is given in [4, Lemma 3]. Indeed in that lemma an external uniform condition is assumed, but it can be checked that exactly the same arguments go through in our situation and lead to the following proposition, which we state without proof.

Proposition 3.5. *Let $(h, u) \in X(e_0; b)$ be a local b-periodic minimizer for the functional F. Then for any $z_0 \in \widetilde{\Gamma}_h^\#$ there exist an orthonormal basis $e_1, e_2 \in \mathbb{R}^2$, and a rectangle*

$$R := \{z_0 + se_1 + te_2 : -a' < s < a',\ -b' < t < b'\},$$

$a', b' > 0$, such that one of the following two representations holds.

(i) *There exists a Lipschitz function $g : (-a', a') \to (-b', b')$ such that $g(0) = 0$ and*

$$\{(x, y) : x \in \mathbb{R} \text{ and } y < h(x)\} \cap R = \{z_0 + se_1 + te_2 : -a' < s < a',\ -b' < t < g(s)\}.$$

Moreover the function g admits left and right derivatives at every point, that are respectively left and right continuous.

(ii) *There exist two Lipschitz functions $g_1, g_2 : [0, a') \to (-b', b')$ such that $g_i(0) = (g_i)'_+(0) = 0$, for $i = 1, 2$, $g_1 \le g_2$, and*

$$\{(x, y) : x \in \mathbb{R} \text{ and } y < h(x)\} \cap \{z_0 + se_1 + te_2 : 0 < s < a',\ -b' < t < b'\} = \{z_0 + se_1 + te_2 : 0 < s < a',\ -b' < t < g_1(s) \text{ or } g_2(s) < t < b'\}.$$

Moreover the functions g_1, g_2 admit left and right derivatives at every point, that are respectively left and right continuous.

Remark 3.6. Note that in case (ii) the point z_0 is either a cusp point or a point in the vertical cut. Moreover, in this case, setting $z_0 = (x_0, y_0)$, the balls of radius ρ_0 centered in $(x_0 \pm \rho_0, y_0)$, are both tangent to $\widetilde{\Gamma}_h^\#$ and lay below $\widetilde{\Gamma}_h^\#$. This observation immediately implies that there are finitely many cusp points and vertical cuts, thus proving assertion (i) in Theorem 2.5.

4 A decay estimate for the gradient of u

Following [6] we are now going to show that given a point in $z_0 \in \widetilde{\Gamma}_h^\# \setminus \left(\Sigma_h^\# \cup \Sigma_{h,c}^\#\right)$, then in a neighborhood of z_0 the integral of $|\nabla u|^2$ over a disk of radius r decays faster than r. As in [6] the key ingredient to get such an estimate is a well known result of Grisvard ([10]) describing the behavior of solutions to the linear elasticity systems at corner points. Here, however, differently from [6], we have to distinguish between the case whether z_0 lies above the x-axis or is on the x-axis. The latter situation is more delicate to handle, while the former will be treated essentially as in [6], with some simplifications.

In this section we will consider open sets whose boundary can be decomposed into three curves

$$\partial\Omega = \Gamma_1 \cup \Gamma_2 \cup \Gamma_3,$$

where Γ_1, Γ_2 are two segments meeting at the origin with an internal angle $\omega \in (0, 2\pi)$ and Γ_3 is a regular curve joining the two remaining endpoints of Γ_1 and Γ_2 in a smooth way. We shall refer to such an open set as to a *regular domain with corner angle* ω. If Ω is a such a domain and $u \in W^{1,2}(\Omega; \mathbb{R}^2)$, we set

$$\sigma(u)[\nu] := \mathbb{C}E(u)[\nu] = \left[\mu(\nabla u + \nabla^T u) + \lambda(\mathrm{div} u)I\right][\nu],$$

where ν is the exterior normal to $\partial\Omega$ and I is the identity map. Moreover, following [10], for any complex number α we denote by S_α a complex valued function given in polar coordinate by

$$S_\alpha := r^\alpha \Phi_\alpha(r, \theta),$$

where Φ_α is a smooth function depending on α and on the corner angle ω. Next result is proven in [10, Theorems I and 6.2].

Theorem 4.1 (Grisvard). *Let Ω be a regular domain with corner and $f \in L^p(\Omega; \mathbb{R}^2)$, $1 < p \leq 2$.*

(a) *Let* $\omega \in (0, 2\pi)$ *and let* $w \in W^{1,2}(\Omega;\mathbb{R}^2)$ *be a weak solution of the Neumann problem*

$$\begin{cases} \mu\Delta w + (\lambda+\mu)\nabla(\mathrm{div} w) = f & \text{in } \Omega \\ \sigma(w)[\nu] = 0 & \text{on } \partial\Omega. \end{cases} \tag{4.1}$$

Then, there exist constants c_α, c'_α *such that*

$$w - \sum c_\alpha S_\alpha - \sum c'_\alpha \frac{\partial S_\alpha}{\partial \alpha} \in W^{2,p}(\Omega;\mathbb{R}^2). \tag{4.2}$$

The first sum is extended to all simple complex roots of equation

$$\sin^2 \alpha\omega = \alpha^2 \sin^2 \omega, \tag{4.3}$$

contained in the strip $0 < \mathrm{Re}\,\alpha < 2 - 2/p$, *while the second sum is extended to all double roots of* (4.3) *contained in the same strip. However,* (4.2) *holds provided that* (4.3) *has no solutions on the line* $\mathrm{Re}\,\alpha = 2 - 2/p$.

(b) *Moreover, if* $\omega \in (0, \pi)$ *and* $w \in W^{1,2}(\Omega;\mathbb{R}^2)$ *is a weak solution of the mixed problem*

$$\begin{cases} \mu\Delta w + (\lambda+\mu)\nabla(\mathrm{div} w) = f & \text{in } \Omega \\ w = 0 \text{ on } \Gamma_1, \sigma(w)[\nu]=0 \text{ on } \Gamma_2 \cup \Gamma_3, \end{cases} \tag{4.4}$$

then (4.2) *holds. In this case the first sum is extended to all simple complex roots of equation*

$$\sin^2 \alpha\omega = \frac{(\lambda+2\mu)^2 - (\lambda+\mu)^2\alpha^2 \sin^2\omega}{(\lambda+\mu)(\lambda+3\mu)}, \tag{4.5}$$

contained in the strip $0 < \mathrm{Re}\,\alpha < 2 - 2/p$, *while the second sum is extended to all double roots of* (4.5) *contained in the same strip. As before,* (4.2) *holds provided that* (4.5) *has no solutions on the line* $\mathrm{Re}\,\alpha = 2 - 2/p$.

Though this deep result gives no information about the roots of equations (4.3) and (4.5), it can be easily proved that the solutions to both equations contained in the strip $0 < \mathrm{Re}\,\alpha < 1$ are bounded. Hence, by analiticity, they are finitely many. A more precise information is provided by the following result, proved in [13, Theorem 2.2].

Theorem 4.2 (Nicaise). *If* $\omega \in (0, 2\pi)$, *then equation* (4.3) *has no root in the strip* $0 < \mathrm{Re}\,\alpha \le \frac{1}{2}$. *Similarly, equation* (4.5) *has no root in the same strip if* $\omega \in (0, \pi)$.

We are now going to combine the two previous results in order to get a useful estimate for the solutions of both problems (4.1) and (4.4). Before proving this estimate we need to recall the following well know Korn's inequality.

Theorem 4.3 (Korn's inequality). *Let $M>0$ and let $\Omega \subset \mathbb{R}^n$ be an open bounded domain starshaped with respect to a given ball $B_r(x_0) \subset \Omega$ and such that* $\operatorname{diam}\Omega \le M$. *Then there exists a constant $C=C(p,N,r,M)>0$ such that for all $u \in W^{1,p}(\Omega;\mathbb{R}^n)$, $p>1$,*

$$\int_\Omega |\nabla u|^p\,dx \le C\left(\int_\Omega |u|^p\,dx + \int_\Omega |E(u)|^p\,dx\right). \tag{4.6}$$

Moreover, if

$$\int_{B_r(x_0)} u\,dx = 0 \quad \textit{and} \quad \int_{B_r(x_0)} (\nabla u - \nabla^T u)\,dx = 0,$$

then (4.6) *holds in the stronger form*

$$\int_\Omega |\nabla u|^p\,dx \le C\int_\Omega |E(u)|^p\,dx. \tag{4.7}$$

We can now pass to the proof of the a priori estimates. To this aim, we recall that an *infinitesimal rigid motion* is an affine displacement of the form $a + Ax$, where A is a skew symmetric 2×2 matrix and a is a constant vector.

Theorem 4.4. *Let Ω be as in Theorem* 4.1. *There exist $p \in (4/3, 2)$ and $C > 0$ such that if $f \in L^p(\Omega;\mathbb{R}^2)$ and $w \in W^{1,2}(\Omega;\mathbb{R}^2)$ is a weak solution to problem* (4.1), *then*

$$\|w\|_{W^{2,p}(\Omega;\mathbb{R}^2)} \le C\big(\|w\|_{L^p(\Omega;\mathbb{R}^2)} + \|f\|_{L^p(\Omega;\mathbb{R}^2)}\big). \tag{4.8}$$

Similarly, if $\omega \in (0,\pi)$ there exist $p \in (4/3,2)$ and C such that for every weak solution to problem (4.4)

$$\|w\|_{W^{2,p}(\Omega;\mathbb{R}^2)} \le C\|f\|_{L^p(\Omega;\mathbb{R}^2)}. \tag{4.9}$$

Proof. We start by proving (4.8).

As already observed, the strip $0 < \operatorname{Re}\alpha < 1$ contains only finitely many solutions to equation (4.3) and from Theorem 4.2 we may conclude that there exists ε such that they are all contained in the strip $\frac{1}{2} + \varepsilon < \operatorname{Re}\alpha < 1$. Therefore, if we choose $p > 4/3$ such that $2 - \frac{2}{p} < \frac{1}{2} + \varepsilon$, from statement (a) of Theorem 4.1 we get that any weak solution to (4.1), with $f \in L^p(\Omega; \mathbb{R}^2)$ is in $W^{2,p}(\Omega; \mathbb{R}^2)$.

To prove the estimate (4.8) let us set $V := \{u \in W^{2,p}(\Omega; \mathbb{R}^2) : \sigma(u)[\nu] = 0 \text{ on } \partial\Omega\}$, $\widetilde{V} := V/\sim$, where for every $u, v \in V$, we have set $u \sim v$ if and only if $u - v$ is an infinitesimal rigid motion. We define a norm in $\widetilde{V}$ setting for every equivalence class $[u]$, with $u \in V$,

$$\|[u]\|_{\widetilde{V}} := \||E(u)|\|_{L^p(\Omega)} + \||\nabla^2 u|\|_{L^p(\Omega)}.$$

Note that this definition is well posed, since if $u \sim v$, then $E(u) = E(v)$ and $\nabla^2 u = \nabla^2 v$. Note also that $\widetilde{V}$ is a Banach space. In fact, if $[u_h]$ is a Cauchy sequence in $\widetilde{V}$, set

$$a_h := \fint_{B_r(z_0)} u_h \, dz, \qquad A_h := \frac{1}{2} \fint_{B_r(z_0)} \left(\nabla u_h - \nabla^T u_h\right) dz,$$
$$v_h := u_h - a_h - A_h(z - z_0),$$

where $B_r(z_0)$ is a fixed ball, such that Ω is starshaped with respect to this ball. Then, from (4.7) and the fact that $\int_{B_r(z_0)} v_h = 0$, it follows easily that the functions v_h converge in $W^{2,p}(\Omega; \mathbb{R}^2)$ to some v. Moreover, from the definition of v_h we have that $\sigma(v_h)[\nu] = 0$ on $\partial\Omega$. Hence, the same is true for v and this proves that $\widetilde{V}$ is a Banach space.

Consider now the operator $L : \widetilde{V} \to L^p(\Omega; \mathbb{R}^2)$ defined for any $[u] \in \widetilde{V}$ as

$$L[u] := \mu \Delta u + (\lambda + \mu)\nabla(\operatorname{div} u).$$

From what we have observed in the first part of the proof it follows that L is a linear, continuous, surjective between two Banach spaces. Therefore to prove (4.8) we only need to show that L is injective. But this is easy since if $L[u] = 0$ for some $u \in V$, then, recalling that $\sigma(u)[\nu] = 0$, we get that

$$\int_\Omega \mathbb{C}E(u) : E(u) \, dz = 0.$$

Therefore, from the coercivity condition (2.4) we deduce that $E(u) = 0$ and thus, Korn's inequality (4.7) yields that $u(z) = a + Az$ for some skew symmetric matrix A, hence $[u] = 0$. From the injectivity of L we then get that there exists $c > 0$ such that for all $u \in V$

$$\||E(u)|\|_{L^p(\Omega)} + \||\nabla^2 u|\|_{L^p(\Omega)} \leq c\|L[u]\|_{L^p(\Omega;\mathbb{R}^2)}.$$

From this inequality and inequality (4.6) estimate (4.8) immediately follows.

The proof of (4.9) is even simpler. As before, from Theorem 4.2 we get that there exists $p > 4/3$ such that any weak solution to (4.4), with $f \in L^p(\Omega;\mathbb{R}^2)$ is in $W^{2,p}(\Omega;\mathbb{R}^2)$. Then, one may introduce the set $V := \{u \in W^{2,p}(\Omega;\mathbb{R}^2) : u = 0 \text{ on } \Gamma_1,\ \sigma(u)[\nu] = 0 \text{ on } \Gamma_2 \cup \Gamma_3\}$ and easily prove that it is a closed subspace of $W^{2,p}(\Omega;\mathbb{R}^2)$. Then, setting $Lu := \mu\Delta u + (\lambda+\mu)\nabla(\operatorname{div} u)$ for every $u \in W^{2,p}(\Omega;\mathbb{R}^2)$, the proof goes on as before. □

The above theorem gives a global estimate of solutions to the linear systems (4.1) and (4.4). However these solutions may have a singularity at the origin, being smooth in the remaining part of Ω. Next result provides the local estimates of the higher order derivatives away from the origin. The proof goes as in [6, Theorem 3.7].

Proposition 4.5. *Let Ω be as in Theorem 4.1 and let $0 < r_0 < r_1$ be such that the open set $A := \{z \in \Omega : r_0 < |z| < r_1\}$ has positive distance from Γ_3. For any integer $k \in \mathbb{N}$ there exists a constant C_k such that, if $w \in W^{1,2}(\Omega;\mathbb{R}^2)$ is a weak solution to either (4.1) or (4.4), then*

$$\sup_A |\nabla^k w|^2 \le C_k \int_\Omega |\nabla w|^2\,dz.$$

Proof. By the Sobolev embedding theorem it is enough to show that for all $k \ge 2$ one has

$$\int_A |\nabla^k w|^2\,dz \le C'_k \int_\Omega |\nabla w|^2\,dz, \tag{4.10}$$

for some constant C'_k independent of w, a weak solution to either (4.1) or (4.4). To this aim let us fix a point $z_0 \in \Gamma_1$ and $r > 0$ such that the ball $B_{2r}(z_0)$ has positive distance both from the origin and Γ_3, and denote by τ and ν the tangential and normal unit vector, respectively. By applying the standard difference quotient argument in the direction τ (note that this argument works under both Neumann and Dirichlet conditions) and then Korn's inequality (4.7), one easily gets that $\frac{\partial w}{\partial\tau}$ belongs $W^{1,2}(\Omega \cap B_{2r}(z_0);\mathbb{R}^2)$ and satisfies the equation

$$\int_{\Omega\cap B_{2r}(z_0)} \mathbb{C}E\Big(\frac{\partial w}{\partial\tau}\Big) : E(\varphi)\,dz = 0$$

for all $\varphi \in W^{1,2}(\Omega \cap B_{2r}(z_0);\mathbb{R}^2)$ vanishing in a neighborhood of $\overline{\Omega} \cap \partial B_{2r}(z_0)$. Thus, choosing in the above equation $\varphi := \eta^2\frac{\partial w}{\partial\tau}$, where η is a smooth cut-off function with compact support in $B_{3r/2}(z_0)$, $\eta \equiv 1$ in $B_r(z_0)$, $\|\nabla\eta\|_\infty \le c/r$, we get from (2.4)

$$\int_{\Omega\cap B_{2r}(z_0)} \eta^2 \left|E\left(\frac{\partial w}{\partial\tau}\right)\right|^2 dz \le c \int_{\Omega\cap B_{2r}(z_0)} |\nabla\eta|^2 |\nabla w|^2 \, dz.$$

Applying (4.7) once more we have

$$\begin{aligned}\int_{\Omega\cap B_r(z_0)} \left|\nabla\left(\frac{\partial w}{\partial\tau}\right)\right|^2 dz &\le \int_{\Omega\cap B_{2r}(z_0)} \left|\nabla\left(\eta\frac{\partial w}{\partial\tau}\right)\right|^2 dz \\ &\le c\int_{\Omega\cap B_{2r}(z_0)} \left|E\left(\eta\frac{\partial w}{\partial\tau}\right)\right|^2 dz \\ &\le c\int_{\Omega\cap B_{2r}(z_0)} |\nabla\eta|^2|\nabla w|^2 \, dz \\ &\le \frac{c}{r^2}\int_{\Omega\cap B_{2r}(z_0)} |\nabla w|^2 \, dz.\end{aligned}$$

Covering $\partial A \cap \Gamma_1$ with a finite number of such balls $B_r(z_0)$, we get that there exist a neighborhood U of $\partial A \cap \Gamma_1$ and a positive constant c such that

$$\int_U \left|\nabla\left(\frac{\partial w}{\partial\tau}\right)\right|^2 dz \le c\int_\Omega |\nabla w|^2 \, dz. \tag{4.11}$$

We have thus estimated the L^2 norm in U of $\frac{\partial^2 w}{\partial\tau^2}$ and $\frac{\partial^2 w}{\partial\tau\partial\nu}$. In order to estimate $\frac{\partial^2 w}{\partial\nu^2}$ we use the Lamé system by rewriting it in terms of $\frac{\partial^2 w}{\partial\tau^2}$ and $\frac{\partial^2 w}{\partial\tau\partial\nu}$. Setting

$$(1,0) = \alpha\tau + \beta\nu, \quad (0,1) = \beta\tau - \alpha\nu,$$

where $\alpha^2 + \beta^2 = 1$, we have

$$\frac{\partial w}{\partial x} = \alpha\frac{\partial w}{\partial\tau} + \beta\frac{\partial w}{\partial\nu}, \quad \frac{\partial w}{\partial y} = \beta\frac{\partial w}{\partial\tau} - \alpha\frac{\partial w}{\partial\nu}$$

and the Lamé system becomes

$$\begin{aligned}&\frac{\partial^2 w_1}{\partial\nu^2}\left[\mu + (\mu+\lambda)\beta^2\right] - \frac{\partial^2 w_2}{\partial\nu^2}(\mu+\lambda)\alpha\beta = f_1, \\ &-\frac{\partial^2 w_1}{\partial\nu^2}(\mu+\lambda)\alpha\beta + \frac{\partial^2 w_2}{\partial\nu^2}\left[\mu + (\mu+\lambda)\alpha^2\right] = f_2,\end{aligned}$$

where f_1 and f_2 are linear combinations of the remaining second order derivatives of w_1 and w_2 with coefficients depending only on λ, μ, α and β. Hence

$$\frac{\partial^2 w_1}{\partial \nu^2} = \frac{f_1\left[\mu + (\mu+\lambda)\,\alpha^2\right] + f_2\,(\mu+\lambda)\,\alpha\beta}{(2\mu+\lambda)\,\mu},$$
$$\frac{\partial^2 w_2}{\partial \nu^2} = \frac{f_2\left[\mu + (\mu+\lambda)\,\beta^2\right] + f_1\,(\mu+\lambda)\,\alpha\beta}{(2\mu+\lambda)\,\mu},$$

which by (4.11) gives

$$\int_U |\nabla^2 w|^2\,dz \le c\int_\Omega |\nabla w|^2\,dz.$$

Since also $\frac{\partial w}{\partial \tau}$ satisfies the Lamé system with the same boundary condition of w on Γ_1 we get that for any $k \ge 2$

$$\int_U |\nabla^k w|^2\,dz \le c_k\int_\Omega |\nabla w|^2\,dz.$$

A similar estimate can be obtained in a neighborhood U' of $\partial A \cup \Gamma_2$ and in a neighborhood of $\overline{A} \setminus (U \cup U')$, thus proving (4.10). □

Next result is a technical lemma concerning the lifting of a normal trace on the boundary of a regular domain with corner to a $W^{2,2}$ function of the whole domain. In the case of Neumann condition it was proved in [6, Lemma 3.12]. However, in order to apply it to mixed Dirichlet-Neumann problems, we also need to treat the case when both the trace of the normal derivative and of the functions are assigned.

Lemma 4.6. *Let Ω be a regular domain with corner angle $\omega \in (0, 2\pi)$.*

(a) *Let $g \in W^{\frac{1}{2},2}(\partial\Omega, \mathbb{R}^2)$ be a function vanishing in a neighborhood of the origin. There exists $v \in W^{2,2}(\Omega, \mathbb{R}^2)$ such that*

$$\sigma(v)[\nu] = \left[\mu(\nabla v + \nabla^T v) + \lambda(\operatorname{div} v)I\right][\nu] = g \qquad \text{on } \partial\Omega$$

and

$$||v||_{W^{2,2}(\Omega;\mathbb{R}^2)} \le c(\Omega)||g||_{W^{\frac{1}{2},2}(\partial\Omega;\mathbb{R}^2)}.$$

(b) *Let $g \in W^{\frac{1}{2},2}(\Gamma_2 \cup \Gamma_3, \mathbb{R}^2)$ be a function vanishing in a neighborhood of the origin.There exists $v \in W^{2,2}(\Omega, \mathbb{R}^2)$ such that*

$$v = e_0(x,0) \quad \text{on } \Gamma_1, \qquad \sigma(v)[\nu] = g \quad \text{on } \Gamma_2 \cup \Gamma_3 \tag{4.12}$$

and

$$||v||_{W^{2,2}(\Omega;\mathbb{R}^2)} \le c(\Omega)\left[||g||_{W^{\frac{1}{2},2}(\Gamma_2\cup\Gamma_3;\mathbb{R}^2)} + |e_0|\right]. \tag{4.13}$$

Proof.
Step 1. It is easily checked that the the condition $\sigma(v)[\nu] = g$ is equivalent to

$$\begin{cases} \dfrac{\partial v_1}{\partial x}(2\mu+\lambda)\nu_1 + \dfrac{\partial v_1}{\partial y}\mu\nu_2 + \dfrac{\partial v_2}{\partial x}\mu\nu_2 + \dfrac{\partial v_2}{\partial y}\lambda\nu_1 = g_1 \\ \dfrac{\partial v_1}{\partial x}\lambda\nu_2 + \dfrac{\partial v_1}{\partial y}\mu\nu_1 + \dfrac{\partial v_2}{\partial x}\mu\nu_1 + \dfrac{\partial v_2}{\partial y}(2\mu+\lambda)\nu_2 = g_2. \end{cases} \tag{4.14}$$

Since

$$\frac{\partial v}{\partial x} = \frac{\partial v}{\partial \nu}\nu_1 - \frac{\partial v}{\partial \tau}\nu_2, \qquad \frac{\partial v}{\partial y} = \frac{\partial v}{\partial \nu}\nu_2 + \frac{\partial v}{\partial \tau}\nu_1,$$

where ν is the exterior normal and τ is the tangent versor to $\partial\Omega$ oriented counterclockwise, system (4.14) can be rewritten as

$$\begin{cases} \dfrac{\partial v_1}{\partial \nu}[\mu + (\mu+\lambda)\nu_1^2] + \dfrac{\partial v_2}{\partial \nu}(\mu+\lambda)\nu_1\nu_2 = \tilde{g_1} \\ \dfrac{\partial v_1}{\partial \nu}(\mu+\lambda)\nu_1\nu_2 + \dfrac{\partial v_2}{\partial \nu}[\mu + (\mu+\lambda)\nu_2^2] = \tilde{g_2}, \end{cases}$$

where

$$\tilde{g_1} = g_1 + \frac{\partial v_1}{\partial \tau}(\mu+\lambda)\nu_1\nu_2 + \frac{\partial v_2}{\partial \tau}(\mu\nu_2^2 - \lambda\nu_1^2) \tag{4.15}$$

and

$$\tilde{g_2} = g_2 + \frac{\partial v_1}{\partial \tau}(\lambda\nu_2^2 - \mu\nu_1^2) - \frac{\partial v_2}{\partial \tau}(\mu+\lambda)\nu_1\nu_2. \tag{4.16}$$

Therefore we have

$$\begin{cases} \dfrac{\partial v_1}{\partial \nu} = \dfrac{\tilde{g_1}[\mu + (\mu+\lambda)\nu_2^2] - \tilde{g_2}(\mu+\lambda)\nu_1\nu_2}{(2\mu+\lambda)\mu} \equiv \dfrac{h_1}{(2\mu+\lambda)\mu} \\ \dfrac{\partial v_2}{\partial \nu} = \dfrac{\tilde{g_2}[\mu + (\mu+\lambda)\nu_1^2] - \tilde{g_1}(\mu+\lambda)\nu_1\nu_2}{(2\mu+\lambda)\mu} \equiv \dfrac{h_2}{(2\mu+\lambda)\mu}. \end{cases} \tag{4.17}$$

Let us now impose that $\frac{\partial v_1}{\partial \tau} = \frac{\partial v_2}{\partial \tau} = 0$. With such a choice, even if ν is discontinuous at the origin, by the assumption on g the functions h_1, h_2 at the right-hand sides of (4.17) are zero in a neighborhood of the origin, hence they are both in $W^{\frac{1}{2},2}(\partial\Omega)$. Therefore we may apply of [9, Theorem 1.5.2.8] to get a function $v \in W^{2,2}(\Omega;\mathbb{R}^2)$ with $v = 0$ and normal derivatives $\frac{\partial v_i}{\partial \nu} = \frac{h_i}{\mu(2\mu+\lambda)}$ on $\partial\Omega$, and such that the $W^{2,2}(\Omega;\mathbb{R}^2)$ norm of v is bounded by a constant times the $W^{\frac{1}{2},2}(\partial\Omega;\mathbb{R}^2)$ norm of (h_1, h_2). Hence assertion (a) follows.

Step 2. We now deal with case (b). Note that, up to a rotation, we may always assume that Γ_1 lies on the positive x semi-axis.

We start by fixing v equal to $e_0(x, 0)$ on the whole boundary $\partial\Omega$. Next, we extend g to a function $\bar{g} : \partial\Omega \mapsto \mathbb{R}^2$. To this aim, if U is a neighbourhood of the origin such that $g \equiv 0$ on $A := U \cap \Gamma_2$, we set $B := U \cap \Gamma_1$ and define $\bar{g}$ as follows:

$$\bar{g}(z) = \begin{cases} (\alpha, \beta) \in \mathbb{R}^2 & \text{if } z \in B \\ g(z) & \text{if } z \in \Gamma_1 \cup \Gamma_3 \\ \text{any smooth interpolation between } (\alpha, \beta) \text{ and } g & \text{otherwise.} \end{cases}$$

We now impose that $\frac{\partial v}{\partial \nu}$ satisfies (4.17) where $\tilde{g}_1$, $\tilde{g}_2$ are defined as in (4.15), (4.16), with g replaced by $\bar{g}$. Then, in order to prove (4.12) it is enough to show that α and β can be chosen in such a way that the functions h_1, h_2 in (4.17) belong to $W^{\frac{1}{2},2}(\partial\Omega)$ and satisfy

$$||h_1||_{W^{\frac{1}{2},2}(\partial\Omega)} + ||h_2||_{W^{\frac{1}{2},2}(\partial\Omega)} \leq c[||g||_{W^{\frac{1}{2},2}(\Gamma_2 \cup \Gamma_3; \mathbb{R}^2)} + |e_0|]. \tag{4.18}$$

Once this inequality is proved, [10, Theorem 1.5.2.8] again will imply that there exists $v \in W^{2,2}(\partial\Omega, \mathbb{R}^2)$, satisfying (4.17) and $v = e_0(x, 0)$ on Γ_1, hence (4.12), such that (4.13) holds.

To prove (4.18) note that

$$\frac{\partial v}{\partial \tau} = \begin{cases} e_0(1, 0) & \text{on } \Gamma_1 \\ e_0(-\nu_2, 0) & \text{on } \Gamma_2. \end{cases} \tag{4.19}$$

In order to choose α, β so that $h_1, h_2 \in W^{\frac{1}{2},2}(\partial\Omega)$ we only need to worry about the regularity of these functions in $A \cup B$. From (4.15), (4.16) and (4.19), we have that

$$h_1 \equiv -2e_0\mu(\mu + \lambda)\nu_1\nu_2^2 \qquad \text{on } A \tag{4.20}$$

while $h_1 \equiv \alpha(2\mu + \lambda)$ on B. Thus, we set α so that $\alpha(2\mu + \lambda)$ is equal to the right hand side of (4.20). Similarly, observing that

$$h_2 = e_0\mu\nu_2[\nu_1^2(2\mu + \lambda) - \lambda\nu_2^2] \qquad \text{on A} \tag{4.21}$$

and that $h_2 = \beta\mu + \lambda\mu e_0$ on B, we choose β so that $\beta\mu + \lambda\mu e_0$ coincides with the right hand-side of (4.21). Note that, with such choice of α and β the functions h_1, h_2 belong to $W^{\frac{1}{2},2}(\partial\Omega)$ (while in general $\bar{g}_1, \bar{g}_2$ do not) and that (4.18) is satisfied. □

We are now in position to prove a decay estimate near the origin for the gradient of solutions either of the Neumann problem (4.1) or of the mixed problem (4.4).

Theorem 4.7. *Let Ω be a regular domain with corner and let $u \in W^{1,2}(\Omega;\mathbb{R}^2)$ be either a weak solution to the Neumann problem*

$$\begin{cases} \mu\Delta u + (\lambda+\mu)\nabla(\mathrm{div} u) = 0 & \text{in } \Omega \\ \sigma(w)[\nu] = 0 & \text{on } \Gamma_1 \cup \Gamma_2, \end{cases}$$

or a weak solution to the mixed problem

$$\begin{cases} \mu\Delta u + (\lambda+\mu)\nabla(\mathrm{div} u) = 0 \qquad \text{in } \Omega \\ u = e_0(x,0) \text{ on } \Gamma_1, \quad \sigma(u)[\nu] = 0 \text{ on } \Gamma_2 \end{cases}$$

(in the latter case assume also $\omega \in (0,\pi)$). Let $r_0 > 0$ such that $\overline{B}_{r_0} \cap \Gamma_3 = \emptyset$. Then, there exist $C > 0$, $\alpha > 1/2$, depending only on λ, μ and Ω, such that for all $r \in (0, r_0)$

$$\int_{B_r\cap\Omega} |\nabla u|^2\,dz \le C r^{2\alpha} \int_\Omega \left(|u|^2 + |\nabla u|^2\right) dz. \tag{4.22}$$

Proof. We shall prove the assertion only for the mixed problem, the other case being similar and actually simpler (see also [6, Theorem 3.11]).

Fix $t \in (0,1)$ so that $\overline{B}_{r_0} \cap t\,\Gamma_3 = \emptyset$ and set $\Omega' := t\,\Omega$. Clearly, Ω' is a regular domain with the same corner angle ω of Ω and boundary $\partial\Omega' = t(\Gamma_1\cup\Gamma_2\cup\Gamma_3)$. By Proposition 4.5 we get that $u \in C^\infty(\Omega'\setminus\overline{B}_{r_1})$, where $0 < r_1 < r_0$, and

$$\int_{\Omega'\setminus\overline{B}_{r_1}} |\nabla^2 u|^2\,dz \le c\int_\Omega |\nabla u|^2\,dz.$$

From this estimate and the assumption $\sigma(u)[\nu] = 0$ on Γ_2, setting $\Gamma := t\,\Gamma_2 \cup t\,\Gamma_3$, we have

$$\|\sigma(u)[\nu]\|^2_{W^{\frac{1}{2},2}(\Gamma;\mathbb{R}^2)} \le c_0 \int_{\Omega'\setminus\overline{B}_{r_1}} \left(|\nabla u|^2 + |\nabla^2 u|^2\right) dz \le C_0 \int_\Omega |\nabla u|^2\,dz,$$

for some constant C_0 independent of u. By applying Lemma 4.6 to the function $g := \sigma(u)[\nu] \in W^{\frac{1}{2},2}(t\Gamma_2 \cup t\Gamma_3;\mathbb{R}^2)$ we get that there exists a function $v \in W^{2,2}(\Omega';\mathbb{R}^2)$ such that

$$v = e_0(x,0) \quad \text{on } t\,\Gamma_1, \qquad \sigma(v)[\nu] = g \quad \text{on } t(\Gamma_2\cup\Gamma_3), \tag{4.23}$$

$$\|v\|_{W^{2,2}(\Omega';\mathbb{R}^2)} \le c\big[\|g\|_{W^{\frac{1}{2},2}(\partial\Omega';\mathbb{R}^2)} + |e_0|\big] \le c\|u\|_{W^{1,2}(\Omega;\mathbb{R}^2)}. \tag{4.24}$$

Defining $w := u - v$, we get from (4.23) that w solves the mixed problem

$$\begin{cases} \mu\Delta w + (\lambda+\mu)\nabla(\mathrm{div} w) = -\mu\Delta v - (\lambda+\mu)\nabla(\mathrm{div} v) \ \text{ in } \Omega' \\ w = 0 \ \text{ on } t\,\Gamma_1, \ \ \sigma(w)[\nu] = 0 \text{ on } t(\Gamma_2 \cup \Gamma_3). \end{cases}$$

Thus, (4.9) implies that

$$\|w\|_{W^{2,p}(\Omega';\mathbb{R}^2)} \leq c\|v\|_{W^{2,p}(\Omega';\mathbb{R}^2)}, \tag{4.25}$$

for some $p > 4/3$ depending only on the Lamé constants and on Ω. Thus, if $0 < r < r_0$, using the Sobolev imbedding theorem, (4.24) and (4.25), we have

$$\begin{aligned} \int_{B_r\cap\Omega} |\nabla u|^2\, dz &\leq 2\int_{B_r\cap\Omega} \left(|\nabla w|^2 + |\nabla v|^2\right) dz \\ &\leq c\left(\int_{B_r\cap\Omega}\left[|\nabla w|^{\frac{2p}{2-p}} + |\nabla v|^{\frac{2p}{2-p}}\right] dz\right)^{\frac{2-p}{p}} r^{\frac{4(p-1)}{p}} \\ &\leq c r^{\frac{4(p-1)}{p}} \left(\|w\|^2_{W^{2,p}(\Omega';\mathbb{R}^2)} + \|v\|^2_{W^{2,p}(\Omega';\mathbb{R}^2)}\right) \\ &\leq c r^{\frac{4(p-1)}{p}} \int_{B_r\cap\Omega} \left(|u|^2 + |\nabla u|^2\right) dz, \end{aligned}$$

hence (4.22) follows with $\alpha := 2(p-1)/p$, which is strictly greater than $1/2$, since $p > 4/3$. □

Let (h, u) be a local minimizer of F. We are now going to describe the behavior of ∇u near a point $z_0 = (x_0, y_0) \in \Gamma_h^\# \setminus (\Sigma_h^\# \cup \Sigma_{h,c}^\#)$, showing that the integral of $|\nabla u|^2$ in a ball $B_r(z_0)$ decays faster than r.

In order to prove this decay estimate we need a slightly different version of the Korn inequality whose proof can be found in [6, Theorem 4.2].

Theorem 4.8 (Korn's Inequality in subgraphs of Lipschitz functions). *Let B' be the unit ball in $\mathbb{R}^{n-1}$ and let $h : B' \to [-L, L]$ be a Lipschitz function with* Lip $h \leq L$ *for some $L > 0$. Define*

$$C_h := \{(x', x_n) \in B' \times \mathbb{R} : \ -4L < x_n < h(x')\}.$$

Then there exists a constant C depending only on N, p, and L such that

$$\int_{C_h} |\nabla u|^p\, dx \leq C\left(\int_{C_h} |u|^p dx + \int_{C_h} |E(u)|^p dx\right)$$

for all $u \in W^{1,p}(C_h;\mathbb{R}^n)$, $p>1$. *Moreover for any ball* B *compactly contained in* $B' \times (-4L,-3L)$ *there exists a constant* C_1 *depending only on* N, p, L *and on the radius of* B *such that*

$$\int_{C_h} |\nabla u|^p\,dx \le C_1 \int_{C_h} |E(u)|^p\,dx$$

for all $u \in W^{1,p}(C_h;\mathbb{R}^n)$ *with*

$$\int_B \big(\nabla u - \nabla^T u\big)\,dx = 0.$$

As a consequence of previous result and Proposition 3.5 we have that if $z_0 \in \Gamma_h^\# \setminus (\Sigma_h^\# \cup \Sigma_{h,c}^\#)$ then there exists a neighborhood U of z_0 such that $u \in W^{1,2}(\Omega^\# \cap U;\mathbb{R}^2)$.

Theorem 4.9. *Let* (h,u) *be a a local minimizer of* F *and* $z_0 \in \Gamma_h^\# \setminus (\Sigma_h^\# \cup \Sigma_{h,c}^\#)$. *Then, there exist* $\alpha > 1/2$, $C_0>0$, $r_0>0$, *such that for all* $r \in (0,r_0)$

$$\int_{B_r(z_0)\cap\Omega_h^\#} |\nabla u|^2\,dz \le C_0 r^{2\alpha}. \tag{4.26}$$

Proof. As in [6, Theorem 3.13] we will prove the result by a blow-up argument. We shall discuss only the case $z_0=(x_0,y_0)$ with $y_0=0$, whose proof is similar but more difficult than the proof needed in the case $y_0>0$, which goes exactly as in [6].

We are going to show that there exist $\alpha>1/2$, $C_0>0$, $r_0>0$, such that for all $r\in(0,r_0)$

$$\int_{B_r(z_0)\cap\{x>x_0\}\cap\Omega_h^\#} |\nabla u|^2\,dz \le C_0 r^{2\alpha}. \tag{4.27}$$

Then, combining (4.27) with a similar estimate in $B_r(z_0)\cap\{x<x_0\}\cap\Omega_h^\#$, we get (4.26).

Step 1. We start by assuming first that $h'_+(x_0)<\infty$. We may also assume that h is not identically zero in a right neighborhood of x_0, since otherwise there is nothing to prove. By Proposition 3.5 there exists $r_1>0$ such that $h_{|(x_0,x_0+r_1)}$ is a Lipschitz function with Lipschitz constant L. For any $0<r<r_1$ set

$$T_r(z_0) := \{(x,y)\in(x_0,x_0+r)\times\mathbb{R} : 0<y<h(x)\}.$$

Set also

$$v(x,y) := u(x,y) - e_0(x,0) \qquad \text{for all } (x,y)\in\Omega_h^\#.$$

We claim that there exist $C_1 > 0$, $\beta > 1/2$ such that for all $\tau \in (0, 1/2]$ there exists $r_\tau > 0$ with the property that

$$\int_{T_{\tau r}(z_0)} |\nabla v|^2 \, dz \le C_1 \tau^{2\beta} \int_{T_r(z_0)} \left(1+|\nabla v|^2\right) dz \quad \text{for all } r \in (0, r_\tau). \tag{4.28}$$

We argue by contradiction assuming that (4.28) does not hold, *i.e.*, there exist $\tau_0 \in (0, 1/2]$ and a sequence of radii r_n converging to zero such that

$$\int_{T_{\tau_0 r_n}(z_0)} |\nabla v|^2 \, dz > C_1 \tau_0^{2\beta} \int_{T_{r_n}(z_0)} \left(1+|\nabla v|^2\right) dz \qquad \text{for all } n \in \mathbb{N} \tag{4.29}$$

(C_1 and β will be determined at the end of Step 3).

Define the sets

$$T_n := \frac{1}{r_n}(T_{r_n}(z_0) - z_0)$$
$$= \left\{ (s,t) \in \mathbb{R}^2 : 0 < s < 1, 0 < t < \frac{h(x_0 + r_n s)}{r_n} \right\}.$$

Setting $g_n(s) := h(x_0 + r_n s)/r_n$, $g_\infty(s) := h'_+(x_0)s$ for $s \in [0, 1]$, since by Proposition 3.5 h'_+ is right continuous in 0, it is easily checked that

$$g_n \to g_\infty \text{ in } C([0,1]), \quad g'_n \to g'_\infty \text{ in } L^p(0,1) \text{ for all } p \ge 1. \tag{4.30}$$

In particular, χ_{T_n} converges almost everywhere to χ_{T_∞}, where

$$T_\infty := \{(s,t) \in \mathbb{R}^2 : 0 < s < 1, \ 0 < t < g_\infty(s)\}.$$

Note that $T_\infty = \emptyset$ if $h'_+(x_0) = 0$. We rescale also v by setting

$$v_n(z) := \frac{v(z_0 + r_n z)}{\lambda_n r_n} \qquad \text{for all } z \in T_n,$$

where $\lambda_n > 0$ and

$$\lambda_n^2 := \frac{1}{|T_{r_n}(z_0)|} \int_{T_{r_n}(z_0)} |\nabla v|^2 \, dz.$$

Up to a (not relabeled) subsequence, we may assume that $\lambda_n \to \lambda_\infty$ as $n \to \infty$. Note that by (4.29) we have that $\lambda_\infty \in (0, \infty]$. Since

$$\frac{1}{|T_n|} \int_{T_n} |\nabla v_n|^2 \, dz = 1, \tag{4.31}$$

and $v_n = 0$ on $\partial T_n \cap \{y = 0\}$, observing that the functions g_n are are Lipschitz continuous with Lipschitz constants bounded by L, we may extend each function v_n to a function (still denoted by v_n) defined in the rectangle $R = (0, 1) \times (0, 2L)$ so that

$$\|v_n\|_{W^{1,2}(R;\mathbb{R}^2)} \leq c(L)\||\nabla v_n|\|_{L^2(T_n)} \leq c.$$

Therefore, up to a subsequence, we may also assume that there exists $v_\infty \in W^{1,2}(R; \mathbb{R}^2)$, with $v_\infty = 0$ on $(0, 1) \times \{0\}$, such that $v_n \rightharpoonup v_\infty$ weakly in $W^{1,2}(R; \mathbb{R}^2)$. Moreover, from the definition of v_n we have that for every $\varphi \in C^1(\overline{R}; \mathbb{R}^2)$ vanishing on $[0, 1] \times \{0\}$ and on $\{1\} \times [0, 2L]$

$$\int_{T_n} \mathbb{C}E(v_n) : E(\varphi)\, dz = -\frac{e_0}{\lambda_n} \int_{\partial T_n} \big[(2\mu+\lambda)\nu_1\varphi_1 + \lambda\varphi_2\nu_2\big]\, d\mathcal{H}^1. \quad (4.32)$$

If $h'_+(x_0) > 0$, letting $n \to \infty$ in (4.32), we get

$$\int_{T_\infty} \mathbb{C}E(v_\infty) : E(\varphi)\, dz = -\frac{e_0}{\lambda_\infty} \int_{\partial T_\infty} \big[(2\mu+\lambda)\nu_1\varphi_1 + \lambda\varphi_2\nu_2\big] d\mathcal{H}^1 \quad (4.33)$$

for every $\varphi \in C^1(\overline{R}; \mathbb{R}^2)$ vanishing on $[0, 1] \times \{0\}$ and on $\{1\} \times [0, 2L]$. In fact, the convergence of the left-hand side of (4.32) to the left-hand side of (4.33) is obvious, while the convergence of the right-hand side follows from (4.30) observing that

$$\begin{aligned}
&\lim_{n\to\infty} \int_{\partial T_n} \big[(2\mu + \lambda)\nu_1\varphi_1 + \lambda\varphi_2\nu_2\big] d\mathcal{H}^1 \\
&= \lim_{n\to\infty} \int_0^1 \big[-(2\mu + \lambda)g'_n(s)\varphi_1(s, g_n(s)) + \lambda\varphi_2(s, g_n(s))\big] ds \\
&= \int_0^1 \big[-(2\mu + \lambda)g'_\infty(s)\varphi_1(s, g_\infty(s)) + \lambda\varphi_2(s, g_\infty(s))\big] ds \\
&= \int_{\partial T_\infty} \big[(2\mu + \lambda)\nu_1\varphi_1 + \lambda\varphi_2\nu_2\big] d\mathcal{H}^1.
\end{aligned}$$

Note that if $h'_+(x_0) = 0$, we get no limit equation, since both sides of (4.32) converge to zero.

Step 2. Let us now fix a function $\psi \in C^1(\overline{R}; \mathbb{R}^2)$ vanishing on $\{1\} \times [0, 2L]$.

If $h'_+(x_0) > 0$, from (4.32) and (4.33) we get, setting $\varphi = \psi^2 v_n$ and $\varphi = \psi^2 v_\infty$, respectively,

$$\begin{aligned}
&\int_{T_n} \psi^2 \mathbb{C}E(v_n):E(v_n)\,dz \\
&= -\int_{T_n} \psi \mathbb{C}E(v_n):\big(v_n \otimes \nabla\psi + (v_n \otimes \nabla\psi)^T\big)\,dz \\
&\quad - \frac{e_0}{\lambda_n}\int_0^1 \big[-(2\mu+\lambda)g_n'(s)v_{n,1}(s,g_n(s))\psi_1^2(s,g_n(s)) \\
&\qquad\qquad + \lambda v_{n,2}(s,g_n(s))\psi_2^2(s,g_n(s))\big]ds \\
&\int_{T_\infty} \psi^2\, \mathbb{C}E(v_\infty) : E(v_\infty)\,dz \\
&= -\int_{T_\infty} \psi\, \mathbb{C}E(v_\infty) : \big(v_\infty \otimes \nabla\psi + (v_\infty \otimes \nabla\psi)^T\big)\,dz \\
&\quad - \frac{e_0}{\lambda_\infty}\int_0^1 \big[-(2\mu+\lambda)g_\infty'(s)v_{\infty,1}(s,g_\infty(s))\psi_1^2(s,g_\infty(s)) \\
&\qquad\qquad + \lambda v_{\infty,2}(s,g_\infty(s))\psi_2^2(s,g_\infty(s))\big]ds.
\end{aligned}$$

Therefore, from the two previous equations we may conclude that

$$\lim_{n\to\infty}\int_{T_n} \psi^2\, \mathbb{C}E(v_n) : E(v_n)\,dz = \int_{T_\infty} \psi^2\, \mathbb{C}E(v_\infty) : E(v_\infty)\,dz, \quad (4.34)$$

provided that we show that the right-hands side converge to the corresponding right-hand side. To this aim, note that the convergence

$$\begin{aligned}
&\int_{T_n} \psi\, \mathbb{C}E(v_n) : \big(v_n \otimes \nabla\psi + (v_n \otimes \nabla\psi)^T\big) \\
&\longrightarrow \int_{T_\infty} \psi\, \mathbb{C}E(v_\infty) : \big(v_\infty \otimes \nabla\psi + (v_\infty \otimes \nabla\psi)^T\big)
\end{aligned}$$

is an easy consequence of the weak convergence in $W^{1,2}(R;\mathbb{R}^2)$ of v_n to v_∞. Let us now prove that

$$\begin{aligned}
&\int_0^1 g_n'(s)v_{n,1}(s,g_n(s))\psi_1^2(s,g_n(s))\,ds \\
&\longrightarrow \int_0^1 g_\infty'(s)v_{\infty,1}(s,g_\infty(s))\psi_1^2(s,g_\infty(s))\,ds.
\end{aligned} \quad (4.35)$$

The proof of the convergence

$$\int_0^1 v_{n,2}(s,g_n(s))\psi_2^2(s,g_n(s))\,ds \longrightarrow \int_0^1 v_{n,2}(s,g_\infty(s))\psi_2^2(s,g_\infty(s))\,ds,$$

is similar and actually simpler. To prove (4.35) observe that, since $v_n(s,0) = v_\infty(s,0) = 0$ for $s \in (0,1)$, we have

$$\begin{aligned}
&\lim_{n\to\infty} \int_0^1 g_n'(s) v_{n,1}(s, g_n(s)) \psi_1^2(s, g_n(s))\, ds \\
&= \lim_{n\to\infty} \int_0^1 g_n'(s) \psi_1^2(s, g_n(s)) \int_0^{g_n(s)} \frac{\partial v_{n,1}}{\partial y}(s,t)\, ds dt \\
&= \lim_{n\to\infty} \int_{T_n} g_n'(s) \psi_1^2(s, g_n(s)) \frac{\partial v_{n,1}}{\partial y}(s,t)\, ds dt \\
&= \int_{T_\infty} g_\infty'(s) \psi_1^2(s, g_\infty(s)) \frac{\partial v_{\infty,1}}{\partial y}(s,t)\, ds dt \\
&= \int_0^1 g_\infty'(s) v_{\infty,1}(s, g_\infty(s)) \psi_1^2(s, g_\infty(s))\, ds.
\end{aligned}$$

Now that we have established (4.34) observe that it implies that

$$\lim_{n\to\infty} \int_{T_n} \psi^2\, \mathbb{C}E(v_n - v_\infty) : E(v_n - v_\infty)\, dz = 0$$

for all $\psi \in C^1(\overline{R}; \mathbb{R}^2)$ vanishing on $\{1\} \times [0, 2L]$. Finally, from this last equation, recalling that $v_n = v_\infty = 0$ on $(0,1) \times \{0\}$ and using Theorem 4.8, we conclude that

$$\lim_{n\to\infty} \int_{\tau T_n} |\nabla v_n - \nabla v_\infty|^2\, dz = 0 \qquad \text{for all } 0 < \tau < 1. \tag{4.36}$$

Let us now show that if $h'_+(x_0) = 0$, we have

$$\lim_{n\to\infty} \int_{\tau T_n} |\nabla v_n|^2\, dz = 0 \qquad \text{for all } 0 < \tau < 1. \tag{4.37}$$

In fact, to prove this equality we choose, as before, $\varphi = \psi^2 v_n$ in (4.32), thus getting

$$\begin{aligned}
\int_{T_n} \psi^2 \mathbb{C}E(v_n) : E(v_n)\, dz &= -\int_{T_n} \psi \mathbb{C}E(v_n) : \left(v_n \otimes \nabla\psi + (v_n \otimes \nabla\psi)^T\right) dz \\
&+ \frac{e_0}{\lambda_\infty} \int_0^1 \big[-(2\mu+\lambda) g_n'(s) v_{n,1}(s, g_n(s)) \psi_1^2(s, g_n(s)) \\
&+ \lambda v_{n,2}(s, g_n(s)) \psi_2^2(s, g_n(s))\big] ds.
\end{aligned}$$

Then we conclude that the right-hand side of this equality tends to zero. In fact

$$\lim_{n\to\infty} \int_{T_n} \psi\, \mathbb{C}E(v_n) : \left(v_n \otimes \nabla\psi + (v_n \otimes \nabla\psi)^T\right) dz = 0,$$

since $v_n \to v_\infty$ weakly in $W^{1,2}(R;\mathbb{R}^2)$ and $\chi_{T_n} \to 0$ a.e. in R, while, arguing as before, we have

$$\begin{aligned}\lim_{n\to\infty}\int_0^1 &\big[-(2\mu+\lambda)g_n'(s)v_{n,1}(s,g_n(s))\psi_1^2(s,g_n(s))\\ &\quad+\lambda v_{n,2}(s,g_n(s))\psi_2^2(s,g_n(s))\big]ds\\ =\int_0^1 &\big[-(2\mu+\lambda)g_\infty'(s)v_{\infty,1}(s,g_\infty(s))\psi_1^2(s,g_\infty(s))\\ &\quad+\lambda v_{\infty,2}(s,g_\infty(s))\psi_2^2(s,g_\infty(s))\big]ds=0,\end{aligned}$$

since $g_\infty \equiv 0$ and $v_\infty(s,0)=0$ for $s\in(0,1)$. Therefore, we may conclude that

$$\lim_{n\to\infty}\int_{T_n}\psi^2\,\mathbb{C}E(v_n):E(v_n)\,dz=0,$$

hence, (4.37) follows.

Step 3. Assume first $h_+'(x_0)>0$ and set

$$u_\infty := \frac{e_0}{\lambda_\infty}(x,0)+v_\infty.$$

From (4.33) we have that u_∞ solves the problem

$$\begin{cases}\mu\Delta u_\infty+(\lambda+\mu)\nabla(\operatorname{div}u_\infty)=0 & \text{in } T_\infty\\ u_\infty=\dfrac{e_0}{\lambda_\infty}(x,0) & \text{on } (0,1)\times\{0\},\\ \sigma(u_\infty)[\nu]=0 & \text{on } \{(s,g_\infty(s)) : s\in(0,1)\}.\end{cases}$$

Therefore, from Theorem 4.7 we get that there exist $C>0$, $\beta\in(1/2,1]$, depending only on λ, μ and $h_+'(0)$, but not on e_0, λ_∞, such that

$$\int_{\tau T_\infty}|\nabla u_\infty|^2\,dz\le C\tau^{2\beta}\int_{T_\infty}\big(|u_\infty|^2+|\nabla u_\infty|^2\big)\,dz \qquad \text{for all } 0<\tau\le 1/2.$$

From this inequality, recalling that $v_\infty=0$ on $(0,1)\times\{0\}$, we have in particular that

$$\int_{\tau_0 T_\infty}|\nabla v_\infty|^2\,dz\le C\tau_0^{2\beta}\int_{T_\infty}\Big(\frac{e_0^2}{\lambda_\infty^2}+|\nabla v_\infty|^2\Big)\,dz,$$

for some C again depending only on λ, μ and $h'_+(x_0)$, but not on e_0, λ_∞. Recalling (4.31) and (4.36) we have

$$\int_{\tau_0 T_\infty} |\nabla v_\infty|^2\,dz = \lim_{n\to\infty}\int_{\tau_0 T_n} |\nabla v_n|^2\,dz$$
$$= \lim_{n\to\infty}\frac{1}{r_n^2\lambda_n^2}\int_{T_{\tau_0 r_n}(z_0)} |\nabla v|^2\,dz \leq C_2\tau_0^{2\beta}\lim_{n\to\infty}\left(1+\frac{e_0^2}{\lambda_n^2}\right),$$

for some C_2 independent of e_0 and λ_∞. Therefore, for n large,

$$\int_{T_{\tau_0 r_n}(z_0)} |\nabla v|^2\,dz < 2C_2\tau_0^{2\beta}(r_n^2\lambda_n^2 + r_n^2 e_0^2),$$

hence,

$$\int_{T_{\tau_0 r_n}(z_0)} |\nabla v|^2\,dz < C_3\tau_0^{2\beta}\int_{T_{r_n}(z_0)} (1+|\nabla v|^2)\,dz,$$

for some C_3 depending only on λ, μ, $h'_+(x_0)$, e_0 and L, but not on λ_∞. This inequality contradicts (4.29), if we choose $C_1 > C_3$, thus proving (4.28). Therefore we may conclude that there exists $C_4 > 0$ such that for all $\tau \in (0, 1/2]$ there exists $r_\tau > 0$ with the property that

$$\int_{T_{\tau r}(z_0)} (1+|\nabla u|^2)\,dz \leq C_4\tau^{2\beta}\int_{T_r(z_0)} (1+|\nabla u|^2)\,dz \quad \text{for all } r \in (0, r_\tau].$$

Fix now $\alpha \in (1/2, \beta)$, $\tau \in (0, 1/2]$ such that $C_4\tau^{2\beta} \leq \tau^{2\alpha}$. Iterating the previous inequality we then get

$$\int_{T_{\tau^k r_\tau}(z_0)} (1+|\nabla u|^2)\,dz \leq \tau^{2k\alpha}\int_{T_{\tau^{k-1} r_\tau}(z_0)} (1+|\nabla u|^2)\,dz \qquad \text{for all } k \in \mathbb{N}.$$

From this inequality (4.27) easily follows.

If $h'_+(x_0) = 0$, the argument is easier. In fact, from (4.37) we have

$$\lim_{n\to\infty}\int_{\tau_0 T_n} |\nabla v_n|^2\,dz = 0,$$

hence

$$\lim_{n\to\infty}\frac{1}{r_n^2\lambda_n^2}\int_{T_{\tau_0 r_n}(z_0)} |\nabla v|^2\,dz = 0$$

and the conclusion follows as before.

Step 4. We are left with the case $h'_+(x_0) = +\infty$. In this case, by Proposition 3.5 there exists a rectangle $R = (x_0, x_0+b')\times(0, a')$ such that $\Omega_h^\# \cap R$ lies above the graph of a Lipschitz function $g : [0, a') \to [x_0, x_0 + b')$ with right and left derivatives at every point, which are right and left continuous respectively and such that $g'(0) = 0$. The proof then goes exactly as before. $\square$

5 Proof of Theorem 2.5

By Proposition 3.5, for each point $z_0 \in \Gamma_h^\# \setminus \left(\Sigma_h^\# \cup \Sigma_{h,c}^\#\right)$ there exists a rectangular neighborhood R such that $\Gamma_h^\# \cap R$ is the graph of a Lipschitz function g having right and left derivatives at every point, which are right and left continuous, respectively. We shall say that z_0 is a *corner point* if the corresponding right and left derivatives of g are not equal. Next result shows that $\Gamma_h^\# \setminus \left(\Sigma_h^\# \cup \Sigma_{h,c}^\#\right)$ has not corner points, hence it is of class C^1.

Proposition 5.1. *Let (h, u) be a local minimizer of F. Then, $\Gamma_h^\# \setminus \left(\Sigma_h^\# \cup \Sigma_{h,c}^\#\right)$ is of class C^1.*

Proof. As we have observed above, it is enough to show that if $z_0 = (x_0, y_0) \in \Gamma_h^\# \setminus \left(\Sigma_h^\# \cup \Sigma_{h,c}^\#\right)$, then z_0 is not a corner point. To this aim we assume $y_0 = 0$, the case $y_0 > 0$ being similar and actually simpler (see [6, Theorem 3.14]).

Let $r \in (0, r_0)$, where r_0 is as in Theorem 4.9, and let $z_r' = (x_r', y_r')$ the point in $\Gamma_h^\# \cap \partial B_r(z_0)$, with the smallest abscissa. Similarly, let $z_r'' = (x_r'', y_r'')$ the point in $\Gamma_h^\# \cap \partial B_r(z_0)$, with the largest abscissa. Set

$$\tilde{h}(x) = \begin{cases} h(x) & \text{if } x \notin [x_r', x_r''] \\ s(x) & \text{if } x \in [x_r', x_r''], \end{cases}$$

where $s : [x_r', x_r''] \to \mathbb{R}$ is the affine function whose graph is the segment connecting z_r' and z_r''. Set $\widetilde{\Omega}_h^\# = \Omega_h^\# \cup \left(\mathbb{R} \times (-\infty, 0]\right)$ and extend u to $\widetilde{\Omega}_h^\#$ so that the new function, still denoted by u, satisfies $u(x, y) = e_0(x, 0)$ for all $x \in \mathbb{R}$, $y \leq 0$. From (4.26) we have that

$$\int_{B_r(z_0) \cap \widetilde{\Omega}_h^\#} |\nabla u|^2 \, dz \leq C r^{2\alpha}$$

for some $\alpha > 1/2$ and $C > 0$ independent of r. Then, since $\widetilde{\Omega}_h^\# \cap B_r(z_0)$ is a Lipschitz domain, we may extend u to a function defined on $B_r(z_0)$ and still denoted by u such that

$$\int_{B_r(z_0)} |\nabla u|^2 \, dz \leq C' r^{2\alpha}$$

where C' is independent of r. Then, from Proposition 3.1 we have

$$\begin{aligned} 0 &\geq F(h, u) - F(\tilde{h}, u) - \Lambda \big| |\Omega_{\tilde{h}}| - |\Omega_h| \big| \\ &\geq \mathcal{H}^1(\Gamma_h^\# \cap B_r(z_0)) - |z_r' - z_r''| - \int_{B_r(z_0)} W(E(u)) \, dz - \Lambda \pi r^2 \\ &\geq |z_r' - z_0| + |z_r'' - z_0| - |z_r' - z_r''| - C' r^{2\alpha} - \Lambda \pi r^2. \end{aligned}$$

Then, from the previous chain of inequalities we obtain that

$$2r - |z_r' - z_r''| \leq cr^{2\alpha},$$

for some $c > 0$ independent of r. Therefore, dividing both sides of the inequality above by r, we have

$$2 - \frac{|z_r' - z_r''|}{r} \leq cr^{2\alpha-1}.$$

Letting $r \to 0$, since $\alpha > 1/2$, we obtain $\lim_{r\to 0^+} \frac{|z_r' - z_r''|}{r} = 2$, thus showing that the left and right tangent lines at z_0 coincide. This concludes the proof. □

Let us now prove that $\Gamma_h^\# \setminus \left(\Sigma_h^\# \cup \Sigma_{h,c}^\#\right) \cap \{y > 0\}$ is of class $C^{1,\alpha}$ for all $\alpha \in (0, 1/2)$. For the sake of completeness, we reproduce here the proof given in [6], with some small simplifications. To this aim we recall the following decay estimate, which is more or less classic and that can be obtained arguing as in Theorem 4.9 (see also [6, Theorem 3.16]).

Theorem 5.2. *Let (h, u) be a local minimizer for the functional F. Then for every closed subarc $\Gamma \subset \Gamma_h^\# \setminus \left(\Sigma_h^\# \cup \Sigma_{h,c}^\#\right) \cap \{y > 0\}$ and for every $0 < \sigma < 1$ there exist a constant $C > 0$ and a radius r_0 such that for all $0 < r < R_0$ and for all $z_0 \in \Gamma$*

$$\int_{B_r(z_0)\cap\Omega_h^\#} |\nabla u|^2\, dz \leq Cr^{2\sigma}.$$

We conclude by proving statement (iii) of Theorem 2.5.

Proposition 5.3. *Let (h, u) be a local minimizer of F. Then, $\Gamma_h^\# \setminus \left(\Sigma_h^\# \cup \Sigma_{h,c}^\#\right) \cap \{y > 0\}$ is of class $C^{1,\alpha}$ for all $\alpha \in (0, 1/2)$.*

Proof. Let us fix an open subarc $\Gamma \subset \Gamma_h^\# \setminus \left(\Sigma_h^\# \cup \Sigma_{h,c}^\#\right) \cap \{y > 0\}$ such that Γ is the graph of a C^1 function. To fix the ideas (the general case being similar), let us assume that $\Gamma = \{(x, h(x)) : x \in (a', b')\}$ and that $h \in C^1([a', b'])$. Set $M := \max\{|h'(x)| : x \in [a', b']\}$ and fix $\sigma \in (1/2, 1)$. Then, from Theorem 5.2 there exist $r_1 > 0$ and $C > 0$ such that for all $x_0 \in (a', b')$ and $r \in (0, r_0)$

$$\int_{C_r(x_0)\cap\Omega_h^\#} |\nabla u|^2\, dz \leq Cr^{2\sigma},$$

where $C_r(x_0) := (x_0 - r, x_0 + r) \times (h(x_0) - Mr, h(x_0) + Mr)$. As in the proof of Proposition 5.1 we may extend u to $C_r(x_0)$ so that the resulting extension, still denoted by u, satisfies the estimate (see Theorem 5.2)

$$\int_{C_r(z_0)} |\nabla u|^2 \, dz \leq C r^{2\sigma}, \tag{5.1}$$

for some C independent of x_0 and r. We set now

$$\tilde{h}(x) = \begin{cases} h(x) & \text{if } x \notin [x_0, x_0 + r] \\ s(x) & \text{if } x \in [x_0, x_0 + r], \end{cases}$$

where $s : [x_0, x_0 + r] \to \mathbb{R}$ is the piecewise affine function whose graph connects $(x_0, h(x_0))$ to $(x_0 + r, h(x_0 + r))$. Then, from Proposition 3.1 and (5.1), arguing as in the proof of Proposition 5.1, we easily get that, for r is sufficiently small (but not depending on x_0),

$$\int_{x_0}^{x_0+r} \sqrt{1 + h'^2} \, dx - \sqrt{(h(x_0 + r) - h(x_0))^2 + r^2} \leq c r^{2\sigma}. \tag{5.2}$$

Using the elementary inequality

$$\sqrt{1 + b^2} - \sqrt{1 + a^2} \geq \frac{a(b - a)}{\sqrt{1 + a^2}} + \frac{(b - a)^2}{2(1 + \max\{a^2, b^2\})^{3/2}}$$

with $a := \fint_{x_0}^{x_0+r} h' \, dx$ and $b := h'(x)$, and integrating the result in $(x_0, x_0 + r)$, we get

$$\begin{aligned} &\frac{1}{2(1 + M^2)^{3/2}} \fint_{x_0}^{x_0+r} \left(h'(x) - \fint_{x_0}^{x_0+r} h' \, ds \right)^2 dx \\ &\leq \frac{1}{r} \int_{x_0}^{x_0+r} \sqrt{1 + h'^2} \, dx - \frac{1}{r} \sqrt{(h(x_0 + r) - h(x_0))^2 + r^2} \leq c r^{2\sigma - 1}, \end{aligned}$$

where we also used (5.2). Thus, in particular,

$$\fint_{x_0}^{x_0+r} \left| h'(x) - \fint_{x_0}^{x_0+r} h' \, ds \right| dx \leq c r^{\sigma - \frac{1}{2}}.$$

A similar inequality holds also in the interval $(x_0 - r, x_0)$. Hence, by [1, Theorem 7.51] we conclude that h is in $C^{1,\sigma-\frac{1}{2}}([a', b'])$. This proves the assertion. □

References

[1] L. AMBROSIO, N. FUSCO and D. PALLARA, "Functions of Bounded Variation and Free Discontinuity Problems", Oxford Mathematical Monographs. The Clarendon Press, Oxford University Press, New York, 2000.

[2] R. J. ASARO and W. A. TILLER, *Interface morphology development during stress corrosion cracking: Part I: Via surface diffusion*, Metall. Trans. **3** (1972), 1789–1796.

[3] E. BONNETIER and A. CHAMBOLLE, *Computing the equilibrium configuration of epitaxially strained crystalline films*, SIAM J. Appl. Math. **62** (2002), 1093–1121.

[4] A. CHAMBOLLE and C. J. LARSEN, *C^∞ regularity of the free boundary for a two-dimensional optimal compliance problem*, Calc. Var. Partial Differential Equations **18** (2003), 77–94.

[5] I. FONSECA, N. FUSCO, G. LEONI and V. MILLOT, *Material voids for anisotropic surface energies*, J. Math. Pures Appl. **96** (2011), 591–639.

[6] I. FONSECA, N. FUSCO, G. LEONI and M. MORINI, *Equilibrium configurations of epitaxially strained crystalline films: existence and regularity results*, Arch. Rational Mech. Anal. **186** (2007), 477–537.

[7] N. FUSCO and M. MORINI, *Equilibrium configurations of epitaxially strained elastic films: second order minimality conditions and qualitative properties of solutions*, Arch. Rational Mech. Anal. (2011), published on line.

[8] M. A. GRINFELD, *Instability of the separation boundary between a non-hydrostatically stressed elastic body and a melt*, Soviet Physics Doklady **31** (1986), 831–834.

[9] P. GRISVARD, "Elliptic problems in nonsmooth domains", Monographs and Studies in Mathematics, 24. Pitman (Advanced Publishing Program), Boston, MA, 1985.

[10] P. GRISVARD, *Singularités en elasticité*, Arch. Rational Mech. Anal. **107** (1989), 157–180.

[11] H. KOCH, G. LEONI and M. MORINI, *On Optimal regularity of Free Boundary Problems and a Conjecture of De Giorgi*, Comm. Pure Applied Math. **58** (2005), 1051–1076.

[12] R. V. KUKTA and L. B. FREUND, *Minimum energy configurations of epitaxial material clusters on a lattice-mismathched substrate*, J. Mech. Phys. Solids **45** (1997), 1835-1860.

[13] S. NICAISE, *About the Lamé system in a polygonal or a polyhedral domain and a coupled problem between the Lamé system and the*

plate equation. I. Regularity of the solutions, Ann. Scuola Norm. Sup. Pisa Cl. Sci. (4) **196** (1992), 327–361.

[14] B. J. SPENCER, *Asymptotic derivation of the glued-wetting-layer model and contact-angle condition for Stranski-Krastanow islands*, Physical Review B **59** (1999), 2011–2017.

[15] B. J. SPENCER and J. TERSOFF, *Equilibrium shapes and properties of epitaxially strained islands*, Physical Review Letters **79** (1997), 4858–4861.

CRM Series
Publications by the Ennio De Giorgi Mathematical Research Center Pisa

The Ennio De Giorgi Mathematical Research Center in Pisa, Italy, was established in 2001 and organizes research periods focusing on specific fields of current interest, including pure mathematics as well as applications in the natural and social sciences like physics, biology, finance and economics. The CRM series publishes volumes originating from these research periods, thus advancing particular areas of mathematics and their application to problems in the industrial and technological arena.

Published volumes

1. Matematica, cultura e società 2004 (2005). ISBN 88-7642-158-0
2. Matematica, cultura e società 2005 (2006). ISBN 88-7642-188-2
3. M. GIAQUINTA, D. MUCCI, *Maps into Manifolds and Currents: Area and* $W^{1,2}$-, $W^{1/2}$-, BV*-Energies*, 2006. ISBN 88-7642-200-5
4. U. ZANNIER (editor), *Diophantine Geometry*. Proceedings, 2005 (2007). ISBN 978-88-7642-206-5
5. G. MÉTIVIER, *Para-Differential Calculus and Applications to the Cauchy Problem for Nonlinear Systems*, 2008. ISBN 978-88-7642-329-1
6. F. GUERRA, N. ROBOTTI, *Ettore Majorana. Aspects of his Scientific and Academic Activity*, 2008. ISBN 978-88-7642-331-4
7. Y. CENSOR, M. JIANG, A. K. LOUISR (editors), *Mathematical Methods in Biomedical Imaging and Intensity-Modulated Radiation Therapy (IMRT)*, 2008. ISBN 978-88-7642-314-7
8. M. ERICSSON, S. MONTANGERO (editors), *Quantum Information and Many Body Quantum systems*. Proceedings, 2007 (2008). ISBN 978-88-7642-307-9
9. M. NOVAGA, G. ORLANDI (editors), *Singularities in Nonlinear Evolution Phenomena and Applications*. Proceedings, 2008 (2009). ISBN 978-88-7642-343-7

Matematica, cultura e società 2006 (2009). ISBN 88-7642-315-4

10. H. Hosni, F. Montagna (editors), *Probability, Uncertainty and Rationality*, 2010. ISBN 978-88-7642-347-5
11. L. Ambrosio (editor), *Optimal Transportation, Geometry and Functional Inequalities*, 2010. ISBN 978-88-7642-373-4
12*. O. Costin, F. Fauvet, F. Menous, D. Sauzin (editors), *Asymptotics in Dynamics, Geometry and PDEs; Generalized Borel Summation*, vol. I, 2011. ISBN 978-88-7642-374-1, e-ISBN 978-88-7642-379-6
12**. O. Costin, F. Fauvet, F. Menous, D. Sauzin (editors), *Asymptotics in Dynamics, Geometry and PDEs; Generalized Borel Summation*, vol. II, 2011. ISBN 978-88-7642-376-5, e-ISBN 978-88-7642-377-2
13. G. Mingione (editor), *Topics in Modern Regularity Theory*, 2011. ISBN 978-88-7642-426-7, e-ISBN 978-88-7642-427-4

Volumes published earlier

Dynamical Systems. Proceedings, 2002 (2003)
Part I: *Hamiltonian Systems and Celestial Mechanics.*
ISBN 978-88-7642-259-1
Part II: *Topological, Geometrical and Ergodic Properties of Dynamics.*
ISBN 978-88-7642-260-1

Matematica, cultura e società 2003 (2004). ISBN 88-7642-129-7

Ricordando Franco Conti, 2004. ISBN 88-7642-137-8

N.V. Krylov, *Probabilistic Methods of Investigating Interior Smoothness of Harmonic Functions Associated with Degenerate Elliptic Operators,* 2004. ISBN 978-88-7642-261-1

Phase Space Analysis of Partial Differential Equations. Proceedings, vol. I, 2004 (2005). ISBN 978-88-7642-263-1

Phase Space Analysis of Partial Differential Equations. Proceedings, vol. II, 2004 (2005). ISBN 978-88-7642-263-1

Fotocomposizione "CompoMat" Loc. Braccone, 02040 Configni (RI) Italy
Finito di stampare nel mese di gennaio 2012
dalla CSR, Via di Pietralata 157, 00158 Roma